Fundamentals of Chemistry

Fourth Edition
Ralph A. Burns

in the Laboratory

Upper Saddle River, NJ 07458

Prentice
Hall

Project Manager: Kristen Kaiser
Senior Editor: Kent Porter Hamann
Editor-in-Chief: John Challice
Executive Managing Editor: Kathleen Schiaparelli
Assistant Managing Editor: Dinah Thong
Production Editor: James Buckley
Supplement Cover Management/Design: Paul Gourhan
Manufacturing Buyer: Ilene Kahn

*Cover Photo: Dale Chihuly, "Saffron and Red Tower," 2000. Glass sculpture. Rutherford,
California. Photo by Mark McDonnell*

© 2003 by Pearson Education, Inc.
Pearson Education, Inc.
Upper Saddle River, NJ 07458

All rights reserved. No part of this book may be reproduced in any form or
by any means, without permission in writing from the publisher.

The author and publisher of this book have used their best efforts in
preparing this book. These efforts include the development, research, and
testing of the theories and programs to determine their effectiveness. The
author and publisher make no warranty of any kind, expressed or implied,
with regard to these programs or the documentation contained in this
book. The author and publisher shall not be liable in any event for
incidental or consequential damages in connection with, or arising out of,
the furnishing, performance, or use of these programs.

Printed in the United States of America

ISBN 0-13-033726-9

Pearson Education Ltd., *London*
Pearson Education Australia Pty. Ltd., *Sydney*
Pearson Education Singapore, Pte. Ltd.
Pearson Education North Asia Ltd., *Hong Kong*
Pearson Education Canada, Inc., *Toronto*
Pearson Educacíon de Mexico, S.A. de C.V.
Pearson Education—Japan, *Tokyo*
Pearson Education Malaysia, Pte. Ltd.
Pearson Education, *Upper Saddle River, New Jersey*

CONTENTS

Introduction

Chemistry, Measurements, and Characteristic Properties

Atoms and Molecules

Chemical Reactions and Quantitative Chemistry

Gases

Energy Changes

Acid and Bases

Organic Chemistry

Laboratory Safety in Chemistry Labs

1. Wear approved safety goggles when you are doing laboratory investigations.

2. Know where fire extinguishers are, their types, and how to use them.

3. Know where the fire blanket is and how to use it.

4. Know where the water eye wash spray is and how to use it.

5. Know where the safety shower is and how to turn it on.

6. Wear shoes in lab.

7. Do not climb or sit on lab desks.

8. Hang coats and umbrellas in appropriate places. Don't forget them.

9. Put books you are not using out of the way, away from water and chemicals.

10. Do not smoke, eat, or drink while in the laboratory.

11. Point test tubes away from any person.

12. Do not inhale chemical fumes.

13. Use the fume hood when noxious fumes are expected.

14. Use tongs or other approved means of handling hot containers.

15. Do not look into the top of hot containers.

16. Set hot containers on ceramic pads.

17. Do not force rubber against glass or metal without lubrications to protect your hands.

18. Report all injuries to the instructor.

19. Do not engage in "practical jokes" or unauthorized experiments.

20. Clean up breakage or spillage immediately.

21. Learn and use proper laboratory techniques.

22. Follow the prescribed lab procedure.

23. Leave lab desk clean and dry.

24. Lock the drawer, shut off gas, and push your chair up to the lab bench when you leave.

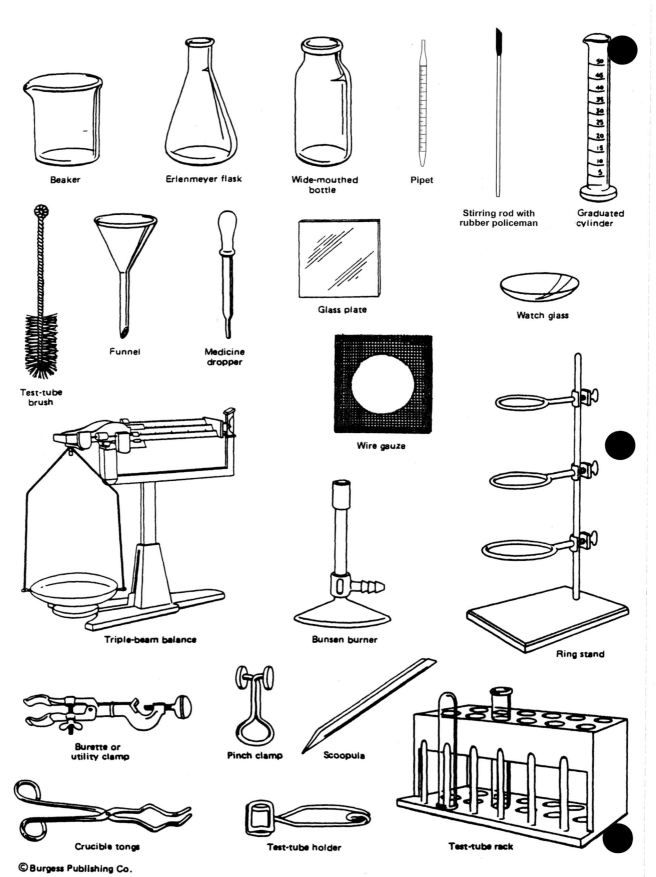

Beaker

Erlenmeyer flask

Wide-mouthed bottle

Pipet

Stirring rod with rubber policeman

Graduated cylinder

Test-tube brush

Funnel

Medicine dropper

Glass plate

Watch glass

Wire gauze

Triple-beam balance

Bunsen burner

Ring stand

Burette or utility clamp

Pinch clamp

Scoopula

Test-tube rack

Crucible tongs

Test-tube holder

© Burgess Publishing Co.

BASIC CHEMISTRY LABORATORY EQUIPMENT

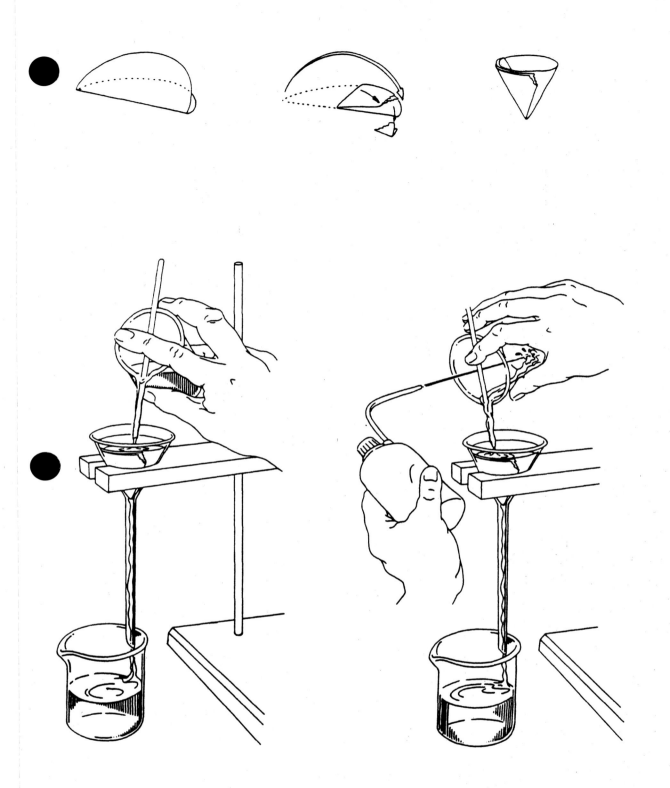

FOLDING FILTER PAPER, USE OF FUNNEL AND WASH BOTTLE

HANDLING REAGENT BOTTLE AND POURING LIQUIDS

FORMAT FOR WRITTEN LABORATORY REPORTS

To be turned in at the end of the session, or at the next laboratory session as your instructor directs. The report should be in ink, or typewritten.

Number of experiment and title (centered on page).

Purpose: This is a single sentence that states what is being investigated and what information the investigations should reveal.

Materials: List only major items necessary for the experiment, omitting common supplies like matches and Bunsen burner.

Procedure: A summary of the general procedure described in the lab manual. This is to be in your own words, not copied. Do not include laboratory observations in this section. Write this in past tense and in third person.

Observations: Data and Results: Compile the data in an organized fashion. Use tables and graphs when appropriate to summarize and compare the findings.

Calculations: Many experiments require at least some calculations. If these are made, show the way you set up the problem and the steps used to get the answer. Do not include the arithmetic of long division or multiplication in long-hand form, just show the mathematical operation performed and the answer. Be certain to include all units, since these are just as important as the numerical portion of the answer.

Questions and Problems: If these are included at the end of an experiment in the lab manual, answer them. Do not put down an answer without first stating the question or problem. These need not be copied directly just summarize them in your own words.

Conclusion: This portion of the report tells what you have learned in the experiment or what values you have determined. Summarize main results in tables. If you were given an unknown, be sure to include its number and your analysis. Consider the purpose of the experiment. Did the investigation give you some pertinent information? Discuss this in detail. How does it relate to what you already know? How reliable is the data? What are some sources of error? Were these significant? Could additional investigations be made? Discuss them.

PHYSICAL AND CHEMICAL CHANGES 1

PURPOSE

To see examples of physical changes and chemical changes and to learn to classify a change as being either physical or chemical.

PERFORMANCE OBJECTIVE

Following these investigations you should be able to describe differences between physical properties and physical changes, and chemical properties and chemical changes. You should also be able to identify examples of each.

BACKGROUND EXPLANATION

Matter frequently undergoes changes. Each change is either a physical change or a chemical change. A physical change involves a change in the form—the size, the state (solid, liquid, or gas), or the structure—of matter. A chemical change always involves a change in the composition of matter; that is, one or more substances are used up while one or more new substances are formed.

If table sugar is heated, it melts to form a colorless liquid. After the liquid cools, it becomes solid and retains the properties it had before being heated. Melting and freezing are physical changes; the composition and physical properties of the sugar were not changed; there is no chemical change.

When charcoal is ignited and burned, it glows red and releases enough heat to grill a steak. The powdery gray ash that is produced bears little resemblance to the original black charcoal; their physical and chemical properties are quite different. For example, the ash will not ignite and burn. That is because the composition of the ash is different from the composition of the charcoal. When charcoal burns, it undergoes a chemical change.

PROCEDURE

In this experiment, several laboratory investigations will be performed on substances and the products will be observed. If a new substance is produced, the change was chemical. If a new substance is not produced, properties remain unchanged; the change is physical.

CAUTION: WEAR SAFETY GOGGLES AT ALL TIMES WHILE YOU ARE IN THE LABORATORY.

Part 1

Using a scoopula or forceps, place 8 to 10 pieces of iodine crystals into a 150 mL beaker.

CAUTION: DO NOT PICK UP IODINE CRYSTALS WITH YOUR FINGERS. AVOID BREATHING IODINE VAPOR WHEN YOU HEAT THE IODINE.

Place the beaker on a wire gauze supported by a ringstand and a ring. Fill an evaporating dish three-fourths full with cool water and place it on top of the beaker containing the iodine. Hold the burner in your hand and gently warm the bottom of the beaker. Continue heating until a collection of crystals on the undersurface of the evaporating dish is clearly evident and no crystals remain in the beaker. Allow the apparatus to cool before removing the evaporating dish. Closely observe the shape of the crystals. If available, use a small magnifying glass to view a few crystals.

With a scoopula, scrape the crystals from the evaporating dish onto a watch glass. Transfer a few crystals into a test tube containing 2 to 3 mL of methylene chloride or acetone and observe the color of the solution. In a second test tube, dissolve the same amount of iodine from the original supply of iodine in 2 to 3 mL of methylene chloride or acetone and compare the colors of the two solutions.

Part 2

Examine a piece of magnesium ribbon that is about five-to-seven centimeters (two-to-three inches) in length. Observe its physical properties: color, luster, and flexibility. Bend the metal strip so most of it hangs over the edge of an evaporating dish supported by a ringstand. Ignite the end of the magnesium with a burner.

CAUTION: DO NOT LOOK DIRECTLY AT THE MAGNESIUM WHILE IT IS BURNING.

Compare the color and texture of the ash with the original metal. Can you bend the ash? Can you ignite it?

Part 3

Into two large test tubes, pour 7 mL of dilute sulfuric acid, H_2SO_4. Roll a sample of copper turnings into a marble-sized ball and place the ball in one of the tubes of acid. Roll another sample of copper turnings into a ball and, using tongs, hold it in the outer cone of the burner flame for four minutes. Let it cool a minute, examine its flexibility, texture, and color, then insert this ball into the other test tube containing acid. Allow both tubes to stand for about seven minutes. Observe the colors of both solutions when held against white paper.

Part 4

Dissolve a pea-sized volume of solid lead(II) nitrate in 20 mL of distilled water in a 100 mL beaker and dissolve a similar volume of solid potassium dichromate in 20 mL of distilled water in another beaker. Stir the solutions to make sure all solids are dissolved. Pour one solution into the other and swirl to mix them.

Prepare to filter the mixture. First, fold a circle of filter paper into fourths and place it in a funnel so it is opened up with one thickness pushed to one side and the other three thicknesses pushed to the other side of the funnel. Hold the paper in place and dampen the filter paper with water from a distilled water bottle.

Place one beaker under the funnel and pour the mixture you prepared into the cone of the filter paper. The liquid that passes through the filter paper is called the **filtrate**; the insoluble solid that is retained on the paper is called the **precipitate**. Rinse the precipitate on the filter paper two times with 5 mL of distilled water from a wash bottle. Discard the filtrate. When the precipitate is completely drained, transfer a small amount—about the volume of a pea—into a beaker containing 20 mL of tap water and stir. Then let it set. Does the precipitate dissolve in water? Did the original two chemicals used in this investigation dissolve in water? Compare colors.

Report Sheet ______________________ ______________ ____________ ______
Name Instructor/Section Lab Partner Date

Experiment 1: Physical and Chemical Properties

Part 1:

1. Describe the color changes that occurred as iodine is heated.

2. Compare the color of the crystals obtained from the bottom of the evaporating dish with crystals of iodine from the original supply.

3. Describe the shape of the crystals formed on the bottom of your evaporating dish. Are they round, flat, jagged or what? Sketch a picture of two or three crystals. Are the crystal shapes similar to one another, or are they quite different?

4. How does the color of the solution containing crystals from your evaporating dish compare to the color of the solution containing the original iodine crystals?

5. Did the iodine undergo a physical change or a chemical change?

Part 2:

1. Describe the color, luster, and texture of magnesium metal.

2. Compare the color, luster and texture of the ash with the original magnesium metal. Will the ash burn?

3. Did the magnesium undergo a physical change or a chemical change? Explain.

Part 3:

1. Compare the appearance, texture, and flexibility of copper metal with the appearance, texture, and flexibility of the material remaining after heating the copper.

2. Compare the color of the acid solution containing copper metal with the color of the solution containing the material remaining after heating copper.

3. Do the materials present before and after heating have the same physical properties? Explain.

4. Do the materials present before and after heating have the same chemical properties? Explain.

5. Did heating the copper cause a physical change or a chemical change? Explain.

Part 4:

1. Describe the results of mixing the two solutions, including color change and solubilities in water.

2. Did a physical change or a chemical change occur here? Explain.

3. On what evidence is your choice based?

QUESTIONS: PHYSICAL AND CHEMICAL CHANGES

Give your true (T) or false (F) answer for each of these questions.

1. _____   You should not look directly at burning magnesium.

2. _____   You should always wear goggles when doing chemistry investigations.

3. _____   A precipitate is the solid that sometimes forms when solutions are mixed; it can be collected on a filter paper.

4. _____   The filtrate is the liquid that passes through the filter paper.

5. _____   Both iodine and table sugar (see background explanation) undergo chemical changes when heated.

6. _____   Analysis of materials present after charcoal is burned (see background explanation) and after copper is heated, shows that the changes were physical, not chemical.

Experiment 1: PRE-LAB or POST-LAB QUIZ

Indicate whether each change described is *physical* or *chemical*.

1. A change in the *size* or *state* (solid, liquid, gas) of matter is a _____ change.

2. A change in the *composition* of matter is a _____ change.

3. When table sugar is heated until it melts, then cooled, a _____ change occurs.

4. When charcoal is heated to a high temperature and cooled, a _____ change occurs.

5. A _____ change occurs when a sample of liquid water evaporates to form a gas (water vapor) when left in an open container.

MEASUREMENTS

PURPOSE

To learn how to use instruments for making metric measurements and to derive conversion factors.

PERFORMANCE OBJECTIVE

Following this experiment you should be able to estimate the length of a dollar bill in centimeters, the mass of a dime or quarter in grams, and the volume of a coffee cup or soda bottle (or can) in milliliters.

BACKGROUND EXPLANATION

Gathering facts and organizing them into meaningful relationships is essential to every scientific inquiry and to solving many types of problems. Chemistry is an experimental science that is dependent on gathering and reporting accurate information. In these investigations, you will be asked to gather and report measurements of length, mass, and volume.

MATERIALS

Calculator, meter stick, balance, cook's measuring cup, test tube, pipet, and graduated cylinder.

PROCEDURE

1. Make the following measurements of length:

   a. Measure the distance across the longest edge of this lab book, first in inches then in centimeters. Record these values on your report sheet. Divide the number of centimeters by inches and record the number of centimeters per inch using three significant figures.
   b. Measure the length of a dollar bill in inches and also in centimeters. Record these measurements to three significant figures. Divide the number of centimeters by inches and record the number of centimeters per inch to three significant figures.

c. How do the cm/in. ratios for (la) and (lb) compare to the accepted value of 2.54 cm/in.? Calculate the percent difference between your ratio and the accepted value (2.54 cm/in.) for part la and lb. What do you conclude?

$$\text{Percent difference} = \frac{\text{difference between your value and the accepted value}}{\text{accepted value}} \times 100\%$$

2. Make the following measurements of masses by "weighing" the items listed:

   a. Use a top loading balance to determine the mass of a dime or quarter to the nearest 0.01 g or 0.001g, based on the balance you use and record the mass and kind of coin.
   b. Use an analytical balance to determine the mass of the same dime or quarter to the nearest 0.1 mg, and record the mass, the kind of coin used, and the date. Note: 0.1 mg = 0.0001 g.

3. Make the following measurements of volume.

   a. Carefully fill a cook's measuring cup with tap water to the 2.0 ounce mark and pour the water carefully into a dry 100 mL graduated cylinder. Carefully read and record this volume of water on the data sheet. When determining the volume of water in milliliters, read the bottom of the meniscus and estimate the volume to tenths of the space between the lines; that is, to tenths of a milliliter. Record these values. Obtain the ratio of milliliters to fluid ounces by dividing the volumes of water in milliliters by the corresponding volume in fluid ounces. Give your answer to three significant figures.
   b. How many milliliters of liquid will a test tube hold? Choose a test tube and fill it with water. Empty the water from the test tube into a graduate cylinder and record the volume of the test tube. Also record its approximate length.
   c. Use a pipet to measure out 2.0 mL of water into a small test tube. You may wish to mark this spot on your test tube with tape or a marking pencil for future reference Repeat for a 4.0 mL and 6.0 mL volume.

CHEMICAL CONVERSION PROBLEMS

Length:

   1. Convert five and one-fourth inches to centimeters. (Given: 1 inch = 2.54 cm.)

   2. Convert 125 mm to inches. Give your answer to three significant figures.

Volume:

    3. What is the volume of a rectangular object with measured dimensions of 14.3 cm by 12.2 cm by 1.50 cm? Include the proper number of significant figures in your answer.

    4. Given that 29.6 mL is equivalent to one fluid ounce, calculate the number of milliliters in a 12.0 oz can of soda.

Mass:

    5. A medicine tablet contains 2.5 mg of an antihistamine. Express this mass in grams.

    6. An adult should limit dietary intake of sodium to 2.40 g/day. Express this as mg/day.

Report Sheet ______________________ ______________ ______________ ________
Name | Instructor/Section | Lab Partner | Date

Experiment 2: Measurement

Part 1a.

Length in inches = ______________ in.

Length in centimeters = ______________ cm

Ratio of centimeters to inches = ______________ cm/in.

Part 1b.

Length of dollar bill in inches = ______________ in.

Length of dollar bill in centimeters = ______________ cm

Ratio of centimeters to inches = ______________ cm/in.

Part 1c.

Percent difference for Part la (see formula in 1c) = ______________ %

Percent difference for Part lb = ______________ %

What do you conclude about your two ratios when compared to the accepted value?

____

____

Part 2a.

Kind of coin (quarter, dime...) ______________

Date of coin ______________

Mass of coin to nearest 0.01 g or 0.001 g ______________ g

Part 2b.

Kind of coin ______________

Date of coin ______________

Mass of coin to nearest 0.1 mg ______________ g

Part 3a.

Actual volume of water in fluid ounces =   _____ fl. oz

Actual volume of water in milliliters =   _____ mL

Ratio of milliliters to fluid ounces =   _____ mL/fl. oz

Part 3b.

Volume of test tube used =   _____ mL

Length of test tube used =   _____ mm

CHEMICAL CONVERSION PROBLEMS

1. Work:

   Answer: _____

2. Work:

   Answer: _____

3. Work:

   Answer: _____

4. Work:

   Answer: _____

5. Work:

   Answer: _____

6. Work:

   Answer: _____

Experiment 2: PRE-LAB or POST-LAB QUIZ

Indicate whether each measurement is of *length*, *volume*, or *mass*.

    1. 11.4 cm is a measurement of _____ .

    2. 8.0 cm^3 is a measurement of _____ .

    3. 24.7 mL is a measurement of _____ .

    4. 725 mg is a measurement of _____ .

Work the following problems:

    5. The sum of 6.0 cm + 2.00 mm + 5.00 mm = _____ cm.

    6. 25 mg = _____ g

    7. A marathon race of 26.2 miles = _____ km.
        Given: 1 mile (mi) = 1760 yd
                   1 m = 1.094 yd
                   1 km = 1000 m

DENSITY **3**

PURPOSE

To learn how to determine densities of solids and liquids.

PERFORMANCE OBJECTIVE

Following these investigations you should be able to use laboratory equipment to obtain densities of solids and liquids.

BACKGROUND EXPLANATION

The density of a substance (D) is defined as the ratio of its mass (m) to its volume (V).

$$D = \frac{m}{V}$$

Density is expressed as mass per unit volume, such as g/cm^3 (read as grams per cubic centimeter) or g/mL (read as grams per milliliter). The slanted line "/" between units means that the first unit (the mass) is divided by the second unit (the volume). The slanted line symbol for division is often used instead of the horizontal division line —— to save vertical space on a page. (See the density formula on this page.)

To experimentally determine the density of a substance, the first step is to obtain the mass of the sample by weighing it. Then, the volume of the sample must be measured. The ratio of mass to volume (the density) is obtained by dividing the mass of the sample by its volume.

MATERIALS

25 mL and 10 mL graduated cylinders, balance, pycnometer, and unknown solid and liquid samples.

PROCEDURE

In this experiment you will be supplied with a piece of metal and a liquid. You are to determine and report their densities. One method of determining the density of a solid and two methods of determining densities of liquids are described here. Your instructor will explain which methods you are to use.

Part 1: The density of a solid

You are to obtain the volume and the mass of your solid metal sample and use these values to calculate density. If a solid has a regular shape such as that of a cube or a cylinder, its volume can be calculated by measuring its appropriate dimensions and calculating the volume. However, if the solid has an irregular shape, its volume can be obtained by submerging it in a measured volume of liquid. The change in volume—the volume of liquid displaced by the solid—is equal to the volume of the metal.

Obtain a metal sample and record the number or letter used to identify the sample. Determine its mass by weighing it to the nearest 0.01 g or 0.001 g and record this mass on your data sheet.

Fill a graduated cylinder (the smallest one the metal will fit in) about two-thirds full of water and, with your eye in line with the bottom of the meniscus, determine the volume of water in the cylinder as precisely as possible. For a graduated cylinder that shows marks at 1 mL intervals, estimate the volume to the nearest 0.1 mL. (When marks are shown at 0.1 mL intervals, estimate the volume to the nearest 0.01 mL.)

Tilt the cylinder and carefully slide the solid sample into the cylinder. Return the cylinder to the upright position and precisely read the new volume at the meniscus. Record this volume on your data sheet. Determine the volume of water displaced by subtracting the original volume of water from the total volume of water and sample. The volume of water displaced in milliliters is equivalent to the volume of the unknown solid sample in cubic centimeters, cm^3.

Determine the density of the metal sample by dividing the mass of the sample by its volume. Record your calculated density in g/cm^3. Finally, compare the density obtained experimentally with values listed in the following table of densities of some common metals to identify your metal sample.

Table of Densities of Some Metals

Metal	Density	Metal	Density
Magnesium, Mg	1.7 g/cm^3	Copper, Cu	9.0 g/cm^3
Aluminum, Al	2.7 g/cm^3	Bismuth	9.7 g/cm^3
Zinc, Zn (dull)	7.1 g/cm^3	Wood's metal (a cadmium and	
Tin, Sn (shines)	7.3 g/cm^3	bismuth alloy)	9.8 g/cm^3
Iron, Fe, or steel	7.8 g/cm^3	Silver, Ag	10.5 g/cm^3
Cadmium, Cd	8.7 g/cm^3	Lead, Pb	11.4 g/cm^3
Nickel, Ni	8.9 g/cm^3	Gold, Au	19.3 g/cm^3

Part 2: The density of a liquid, pycnometer method

Determine the mass of a dry, empty pycnometer to the nearest 0.0001 g by weighing it on an analytical balance. Record this mass. Fill the pycnometer to overflowing with the liquid sample and insert the stopper, letting excess liquid overflow from the tiny hole in the top of the stopper. Wipe excess liquid from the pycnometer and weigh the pycnometer filled with the liquid sample. The mass of the liquid sample is the difference between these two masses.

Empty the pycnometer, rinse it, refill it with distilled water, and weigh it. This mass minus the mass of the empty pycnometer gives the mass of distilled water in the pycnometer. The mass of distilled water, in grams, is numerically equivalent to the volume of the pycnometer and the volume of the unknown sample, in milliliters. That is, 1.000 g of water has a volume of 1.000 mL when the density of water is 1.000 g/mL.

The density of the unknown liquid is calculated by dividing the mass of the liquid sample by its volume.

Part 3: The density of a liquid and percent composition, graphing method

In this investigation, you will determine the density of an unknown salt solution and determine the percent of salt in the solution (the salt concentration) by using a graph of densities and concentrations for solutions with known concentrations.

Weigh a clean, dry 25-mL graduated cylinder as precisely as possible and record its mass. Add about 20 mL of distilled water and record the volume to the nearest 0.1 mL. Weigh the cylinder with the water and record this mass on the data sheet. Determine the mass of the water by subtracting the mass of the dry cylinder from the mass of the cylinder with the water. Calculate the density of the water.

Pour about 20 mL of 4.0% sodium chloride solution (4.0% NaCl by mass) in a clean, dry beaker as your supply. Rinse your graduated cylinder with about 4 to 5 mL of the 4.0% NaCl solution, and pour between 10 and 15 mL of the 4.0% NaCl solution into your graduated cylinder. Determine the volume of the solution in the cylinder as precisely as possible (estimate volume to the nearest 0.1 mL) and record this volume. Weigh the cylinder with this salt solution and calculate the density by dividing the mass of the salt solution by its volume.

Empty the cylinder and beaker and hold them upside down to drain for at least 15 seconds. Repeat the above procedure to determine the densities of the 8%, 12%, and 16% NaCl solutions. (Remember to rinse the graduated cylinder with each new solution to be used.) Calculate the density of each solution.

Finally, obtain approximately 20 mL of the salt solution whose concentration is to be determined, and record its number. Determine its density using the above method and record this density.

To determine the percentage concentration of this salt solution, you will need to plot a graph of salt solution density versus NaCl concentrations.

Use the following steps to prepare the graph.

PLOTTING THE GRAPH

1) Plan to plot the densities on the vertical axis and the concentrations (as percentages) on the horizontal axis. The completed graph should fill most of the page.
2) The first step is to determine the scale, that is, how many units each division on your graph will represent. Values for densities will range from about 1.00 g/mL to 1.30 g/mL. Assuming that your graph paper has about 90 divisions (squares) in the vertical direction, you could use three small squares to represent each 0.01 g/mL so each 30 squares would represent a change of 0.10 g/mL.
3) Determine the starting values for both vertical and horizontal axes. Because the density of pure water at room temperature is 1.00 g/mL, that value would be a good

starting point on the vertical axis. It is not necessary to begin every graph at zero, but the values should always increase as you move away from the origin (the corner of the graph where the vertical and horizontal axes come together.) For the horizontal axis, the values will range from 0% NaCl to 16% NaCl, so if you have 64 horizontal divisions on the graph paper, you could use four divisions for 1.0% (or 4 x 16 = 64 divisions for values ranging from 0% to 16%) and have room for margins on each side of the graph.

4) Number—and mark off—the major divisions along each axis.

5) Label each axis with a description and the units, such as "density in g/mL."

6) Plot the points. For example, for the 4.0% NaCl solution, locate that percentage on the horizontal axis and imagine a line extending straight upward from that point to where it would intersect an imaginary line running horizontally at the corresponding density. Put a small—but distinct—dot at the point where these imaginary lines intersect on your graph. Draw a small circle around each dot that is plotted to make it clearly visible.

7) Draw a straight line that runs through the points (or as near as possible to the points). For laboratory data, the best smooth line will not necessarily touch each of the plotted points, due to errors in measurements. Use a ruler or any straight edge to assist you in drawing a straight line. For some types of graphs the best smooth line is a straight line, but for other graphs, the smooth line is a curve.

8) Title the graph. A good title is brief but explicit.

Next, use the graph to determine the concentration of NaCl (the percentage) for your unknown salt solution. From the point on the vertical axis that represents the density of the unknown, draw a horizontal line that extends to the point where it intersects the line drawn through the points. From this intersection, draw a vertical line extending straight down to the horizontal axis. At the point where this vertical line intersects the horizontal axis, read the concentration of the unknown solution. Record this percentage for your unknown salt solution on your data sheet. Be sure to attach your graph to your data sheet before you turn it in.

Experiment 3: Density

Part 1: Density of a Metal

    a. Unknown metal number (or letter)   _____

    b. Mass of metal sample   _____ g

    c. Volume of water in graduated cylinder   _____ mL

    d. Volume of water plus metal sample   _____ mL

    e. Volume of water displaced (volume of metal)   _____ cm^3

    f. Density of metal number _____   _____ g/cm^3

    g. Probable identity of the metal (see table)   _____

Part 2: Density of a liquid, pycnometer method

    a. Unknown liquid number (or letter)   _____

    b. Mass of empty pycnometer   _____ g

    c. Mass of pycnometer plus unknown liquid   _____ g

    d. Mass of unknown liquid (c minus b)   _____ g

    e. Mass of pycnometer filled with pure water   _____ g

    f. Mass of water in pycnometer (___ minus ___)   _____ g

    g. Volume of pycnometer, assuming the density of
        pure water is 1.000 g/mL, so 1 g water = 1 mL   _____ mL

    h. Density of unknown liquid number _____   _____ g/mL

Part 3: Density of a liquid, graphing method

    a. Mass of graduated cylinder   _____ g

    b. Volume of water in cylinder   _____ mL

    c. Mass of cylinder and water   _____ g

d. Mass of water alone _____ g

e. Calculated density of water _____ g/mL

Salt Solutions   **4.0%   8.0%   12.0%   16.0%**

   f. Volume of NaCl solution   ——   ——   ——   ——
     and graduated cylinder

   g. Mass of NaCl solution and   ——   ——   ——   ——
     graduated cylinder

   h. Mass of NaCl solution alone (3g − 3a) —— —— —— ——

   i. Density of NaCl solution (3h/3f)   ——   ——   ——   ——

Unknown NaCl solution number (or letter)

   j. Unknown solution number (or letter) _____

   k. Volume of unknown NaCl solution _____ mL

   l. Mass of cylinder and unknown solution _____ g

   m. Mass of unknown solution, alone _____ g

   n. Density of unknown solution number _____ _____ g/mL

   o. Concentration of unknown solution (from graph) _____ %

DENSITY PROBLEMS TO DO

1. a. Calculate the density of a metal object that has a dull surface and an irregular shape. It has a mass of 321 g and a volume of 45.2 cm^3.

Work:

Answer: _____

   b. Based on your answer, could this object be made of one of the metals listed in Part 1 of this experiment? If so, which one? Explain.

Answer:

Report Sheet ______________________ ______________ __________ ______
Name Instructor/Section Lab Partner Date

2. a. Rearrange the following density equation, D = m/V, and solve for V.

Work:

Answer: ________________

b. What is the volume of 4.00 g of air if the density of air at the time of the measurement is 1.19 g/L?

Work:

Answer: ________________

3. Calculate the mass of a quart (946 mL) of mercury. The density of mercury is 13.6 g/mL.

Work:

Answer: ________________

[Ex. 3] Density

Experiment 3: PRE-LAB or POST-LAB QUIZ

1. The density of a sample can be obtained by dividing its _____ by its

   _____ .

2. What is the density of a gasoline sample that has a volume of 13.3 mL and a

   mass of 8.86 g? _____

3. A sample of 50.0 mL of sulfuric acid that has a density of 1.84 g/mL would have a

   mass of _____ g.

4. Describe how you can use a graduated cylinder to determine the volume of a
   small, irregularly-shaped solid.

QUANTITATIVE SEPARATION OF A MIXTURE

PURPOSE

To separate a mixture of substances using physical—not chemical—changes. In this investigation we will determine the percentages of table salt and sand in a mixture.

PERFORMANCE OBJECTIVES

Following this investigation you should be able to use filtration and evaporation to separate a substance that is water-soluble from one that is not soluble, and then calculate the percentages of the substances in the sample.

BACKGROUND EXPLANATION

The components of a mixture are not chemically bonded to one another, so they can be separated from one another by physical processes. For example, pure table salt can be separated from an aqueous salt solution (i.e. a saltwater solution) by evaporation of the water. Heating causes the water to evaporate, but the salt does not evaporate.

In this experiment you will be given a mixture of table salt, NaCl, and sand, SiO_2. The salt will be extracted from the mixture of salt and sand by dissolving the salt in water. The resulting saltwater solution and sand will be separated by filtration. After evaporating the samples to dryness, the dried salt and sand will be weighed and percentages will be calculated.

PROCEDURE

Obtain a sample of a mixture of salt and sand from your laboratory instructor and record the number (or letter) of your sample on your data sheet. Weigh a clean, dry 150 mL beaker to at least the nearest 0.01 grams and record this mass on the data sheet. Then, pour all of the salt and sand sample (3 to 5 grams) into the beaker and reweigh the beaker with the sample. Record this mass. By subtracting, you can precisely determine the mass of your sample. Also, weigh a clean, dry porcelain evaporating dish and record its mass.

Using your graduated cylinder, measure out 10.0 mL of distilled (or deionized) water and pour it into the beaker containing the salt mixture. Place the beaker on a wire gauze on a ring stand, adjusted so the ring is about 2 to 3 centimeters above the burner flame. Adjust the burner for a small flame and warm the beaker and its contents.

While the beaker and sample are heating gently, obtain a piece of filter paper that has a diameter of about 13 cm. Fold it over twice (to make a 1/4 pie shaped circle), weigh it, and put it in a glass funnel that is supported by a small ring on your ringstand (or use a larger ring with a triangle) over a 250-mL beaker to collect the **filtrate** (the liquid that goes through the filter paper).

When the contents of the beaker have become quite warm—but not boiling—stir the sample with a stirring rod to make sure all of the salt has dissolved. Remove the beaker from the ringstand.

CAUTION: Hold the beaker only by the upper rim which should not be hot.

Pour the contents of the beaker through the filter paper cone and collect the filtrate. With the warm beaker held at an angle so liquid will run out of it, use a wash bottle to spray a minimal amount of distilled water (2 to 3 mL) up into the beaker to rinse out any remaining salt and sand. Pour two 5-mL portions of distilled water through the filter paper to wash remaining traces of salt into the beaker containing the filtrate. Gently remove the filter paper with the sand from the funnel and spread it out on a watch glass to dry under a heat lamp (or in an oven set at about 110 °C). Ten minutes under a heat lamp (or twenty minutes in an oven) should be enough time to dry the sand. When completely dry, weigh the filter paper and sand. Record this mass. While the sand is drying, pour a portion of the filtrate into the preweighed evaporating dish and place it on the wire gauze over a moderate flame. Heat—but *do not* boil—the liquid to evaporate water from the dish. Remove the flame completely from under the evaporating dish if it begins to boil or spatter. Add any remaining filtrate from the beaker and continue to evaporate the liquid to dryness. Allow the evaporating dish to cool for a few minutes before weighing the evaporating dish containing the salt. Record this mass.

After doing the calculations, turn in your samples of salt and sand as well as your report sheet to your instructor.

Experiment 4: SEPARATION OF A MIXTURE

Data

1. Sample number (or letter)   _____

2. Mass of the BEAKER WITH the salt-and-sand SAMPLE   _____ g

3. Mass of the BEAKER alone   _____ g

4. Mass of the salt-and-sand SAMPLE (2 minus 3)   _____ g

5. Mass of the FILTER PAPER and dried SAND   _____ g

6. Mass of the FILTER PAPER alone   _____ g

7. Mass of the DRIED SAND (5 minus 6)   _____ g

8. Mass of the EVAPORATING DISH and the dried SALT   _____ g

9. Mass of the EVAPORATING DISH alone   _____ g

10. Mass of the DRIED SALT (8 minus 9)   _____ g

Calculations

11. TOTAL masses of recovered salt and sand (7 + 10)   _____ g

12. Percent by mass of salt and sand RECOVERED
    from sample number

$$\frac{\text{Total masses of recovered salt and sand, line 11}}{\text{Mass of the original sample, line 4}} \times 100\%$$ _____ % recovered

13. Percent by mass of SAND in sample number

$$\frac{\text{Mass of the dried sand, line 7}}{\text{Mass of the original sample, line 4}} \times 100\%$$ _____ % sand

14. Percent by mass of SALT in sample number

$$\frac{\text{Mass of the dried salt, line 10}}{\text{Mass of the original sample, line 4}} \times 100\%$$ _____ % salt

A MIXTURE PROBLEM TO DO

If a chemist analyzes a 3.84 g sample (number 118) containing sand and table sugar, and recovers 1.43 g of sand, what percent by mass of table sugar should the chemist report for this sample?

Sample 118 = _____ % table sugar

Experiment 4: PRE-LAB or POST-LAB QUIZ

1. If 3.64 g of sand is separated from 5.24 g of a mixture of salt and sand, what is the percentage of SAND in the sample?

2. What is the percentage of SALT in the sample described in Question 1?

3. Multiple choice; circle your answer. Separating salt from sand involves

   a. a physical change.
   b. a chemical change.
   c. both physical and chemical changes.

4. What is a FILTRATE?

MELTING POINT, A CHARACTERISTIC PROPERTY

5

PURPOSE

To identify an unknown compound by its melting point.

PERFORMANCE OBJECTIVES

Following this investigation, you should be familiar with techniques used to determine the melting point of an unknown substance. You should also be able to use available tables listing melting points to narrow the possible identity of an unknown to several choices, and to make a positive identification of the unknown using mixture melting points.

MATERIALS

200°C thermometer, capillary tube, 250 mL beaker, cooking oil, rubber ring cut from latex rubber tubing, ring stand and burner.

BACKGROUND EXPLANATION

The melting point of a substance is one of the characteristic physical properties chemists use to identify a substance. Other characteristic properties include boiling point, density, color, surface tension, viscosity (rate of flow), and chemical reactivity.

The melting point is the temperature at which the substance changes from the solid to the liquid state. For pure compounds, the melting point is sharp with a range of 0.5 to 1°C. Impure substances do not have a sharp melting point but have a broad melting range that is somewhat lower than that of the pure compound. Therefore, if an unknown is mixed with several suspected compounds, only the mixture with the identical compound will have the same sharp melting point. The other mixtures, being impure, will have lower, broader melting ranges.

PROCEDURE

Use a thin-walled capillary tube that is about 10 cm long and 1-2 mm in diameter and sealed at one end. If the capillary tube you use is not already sealed at one end, seal it at one end by heating the tip of the tube in the edge of a Bunsen burner flame.

Pulverize a small sample of the unknown sample in a watch glass using the end of a large test tube. Place this crushed sample on a watch glass and press the open end of the capillary tube into the crystalline powder. Turn the sealed end of the capillary tube down and repeatedly tap it lightly against the lab bench to let the powder fall into the sealed end to a depth of about 3 mm (1/8 inch). For an alternate method of transferring the powder to the closed end, drop the capillary down a plastic tube—several times—onto the desk top. The plastic tube should be about 3/8 inch by 12 inches, like the plastic case for some laboratory thermometers.

Using a small slice from a piece of latex tubing, attach the capillary to a 200 °C thermometer so the sealed end of the capillary is near the mercury bulb. Suspend the thermometer and the capillary in an oil bath as shown. Use a thermometer clamp or use a cork with a hole and a slit on its side and a utility clamp to hold the thermometer.

Heat the oil bath SLOWLY at a rate of about one degree per minute.* Stir the oil bath continuously. Be sure the burner flame is turned low and that it is positioned properly, with the inner cone about 2-3 cm (1 inch) below the beaker.

The melting point is the temperature at which liquid first appears.

Compare the melting point of the unknown to the values in the table and list the possible identities of your unknown. Your unknown is one of those listed on the accompanying table. Let the temperature of the oil drop about 10 °C and repeat the experiment if time permits.

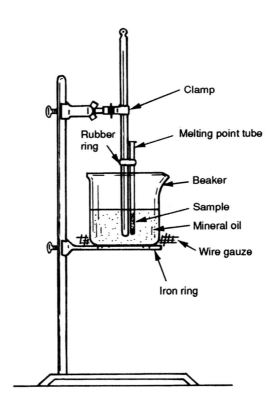

Clamp

Rubber ring

Melting point tube

Beaker

Sample

Mineral oil

Wire gauze

Iron ring

*Time may be saved by taking an approximate melting point with rapid heating, then allowing the oil to cool for a few minutes, before taking the precise melting point.

Report Sheet ______________________ ______________ ______________ ________
Name Instructor/Section Lab Partner Date

Experiment 5: Melting Point, A Characteristic Property

Melting Points of Some Organic Substances in °C	
benzophenone	48–50
bibenzyl	50–51
6-chlorothymol	58–59
palmitic acid	61–63
acetamide	79–81
vanillin	81–83
d,l-mandelic acid	119–121
benzoic acid	121–122
trans-cinnamic acid	133–135
cholesterol	144–146
anthranilic acid	146–148

Data

Unknown number ______________

Melting point (first attempt) ______________ °C

Substances to consider ______________________________

____

Melting point (second attempt) ______________ °C

Conclusion

Unknown number ______________

Melting point ______________ °C

Name of compound (from table) ______________

Discussion: ______________________________

____

Experiment 5: PRE-LAB or POST-LAB QUIZ

Multiple choice questions. Circle your answers.

1. Melting point is

    a. a physical property of a substance.
    b. a chemical property of a substance.
    c. both a physical and a chemical property of a substance.
    d. not a physical or a chemical property of a substance.

2. The melting point (or range) of an IMPURE compound is

    a. the same as the melting point of the pure compound.
    b. higher than the melting point of the pure compound.
    c. lower than the melting point of the pure compound.
    d. sometimes higher and sometimes lower than the melting point of the pure compound.

PAPER CHROMATOGRAPHY

6

PURPOSE

To use paper chromatography to separate and identify five metal cations of the first transition series, Mn^{2+}, Fe^{3+}, Co^{2+}, Ni^{2+} and Cu^{2+}, present in known solutions and in solutions of unknowns.

PERFORMANCE OBJECTIVE:

Following these investigations, you should be able to use paper chromatography to separate and identify cations present in an unknown sample.

BACKGROUND EXPLANATION

Chromatography is a technique used for separating chemical species based on their varying rates of movement by—or through—a medium, usually liquid or gas, along a pathway. Thin layer chromatography utilizes a plate coated with silica gel or a similar pathway. Paper chromatography utilizes absorbent paper as a pathway. Column chromatography utilizes a vertical tube with a granular solid as a pathway. Gas chromatography utilizes a narrow diameter tubing, coated internally, as a pathway. Various liquids, solutions or gases are used to carry the sample. Separation depends on differences in molar masses, polarities of molecules, and other factors.

In this experiment we will use paper chromatography.

PROCEDURE: Wear Eye Protection

Obtain a piece of chromatography paper (such as Whatman No. 1) 20 cm long x 10 cm wide on which to make a chromatogram. Using a pencil, draw a straight line longways 2.0 cm

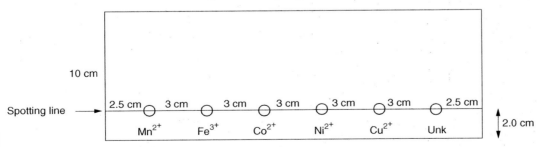

Preparing the Chromatogram

from one 20-cm edge. Mark a small circle—the same size shown on the diagram—on the line 2.5 cm from one end. Then mark similar small circles at 3 cm intervals along this line until 6 circles have been marked. The 6th one should be about 2.5 cm from the other end. These are spotting points.

Spot each cation solution and an unknown solution in this order: Mn^{2+}, Fe^{3+}, Co^{2+}, Ni^{2+}, Cu^{2+}, and unknown—six spots total. For spotting, use a separate small capillary tube for each solution, dip it in the solution and touch it to the marked position on the spotting line. The longer you touch it to the paper the larger the spot will be. A small spot about 3 to 5 mm across (in diameter) is suggested—the same size as shown in the diagram. To intensify the spot, if desired, allow it to dry and then touch the capillary to the same spot a second time.

Allow the spots to dry. Record the color of each dry spot under DATA. One or more of the spots may be colorless because the color intensity of such a thin layer is quite low.

Form the chromatogram into a cylinder with the spots to the outside and staple it near the top and middle. The edges should just touch but NOT overlap. Overlapping changes the conditions and causes the spots near each end to move at an angle toward the middle instead of straight up as desired.

In a 600 mL tall form (no lip) beaker add 34 mL of acetone, 7 mL of 12M HCl (con. HCl) and 4 mL of distilled (deionized) water. Measure volumes using a graduated cylinder and stir the solution with a glass rod. This is the developing solution. Caution: It contains strong acid.

Place the chromatogram (formed into a cylinder) in the beaker containing the developing solution with the spots down. Let the cylinder rest on the bottom of the beaker, but it should not touch the sides of the beaker. Immediately place a glass plate or glass petri dish over the top of the beaker to prevent evaporation of the developing solution, but *do not* move the beaker.

Caution: Do not directly breathe the developing solution—the fumes are irritating.

Allow the developing solvent to soak up into the chromatogram until the solvent front reaches 1 cm from the top or for 30 minutes, whichever comes first. (It may take only 15 minutes.) Remove the chromatogram from the beaker and draw a light pencil line at the solvent front (high solvent mark) so it can be measured when it is dry.

Dry the chromotogram under an infrared heat lamp at 30 cm (12 in.), for 2 to 4 minutes or until dry. Record the colors of each spot on the dry chromatogram under DATA.

Place the chromatogram cylinder in an ammonia chamber (containing concentrated ammonia) in the hood. Do not place the chromatogram directly in the ammonia solution, but permit it to be exposed to the ammonia vapor for 2 or 3 minutes. Remove the chromatogram and observe the colors of the spots. Record the colors of each spot after exposure to ammonia under DATA.

Remove the staples from the the chromatogram cylinder and open it up flat. Dip one end of a double-tipped unused cotton swab (Q-tip) in 0.25M sodium hydroxide, NaOH, and streak it vertically (down to up) through the region where the manganese(II), Mn^{2+}, spot should be found. The previously invisible spot should turn dark brown. Dip the other end of the cotton swab in the 0.25M NaOH and touch the spot in the unknown vertical region where you would expect Mn^{2+} to be if it is present in your unknown. This spot should be at the same distance above the spotting line as the known Mn^{2+}. Do not streak through all of the unknown's vertical region; just contact the place where Mn^{2+} should be. If Mn^{2+} is present what color should it be? Record the color of the Mn^{2+} known and unknown under DATA.

Dip another unused cotton swab in 1% dimethylglyoxime and streak the vertical region where the nickel(II), Ni^{2+}, ion moves. When you contact the previously invisible (or light green) spot with dimethylglyoxime, it should turn bright red. Dip the other end of the swab in 1% dimethylglyoxime and touch the place in the vertical region above the original unknown spotting point where the Ni^{2+} is expected to be; that is, in line with where the known Ni^{2+} appears. Do not streak all the vertical region of the unknown. If Ni^{2+} is present in your unknown what color will it be? Now put the chromatogram back in the NH_3 chamber for 1 min. Double check the colors listed under DATA.

Calculate the R_f value for each of the known cations (in decimals).

$$R_f = \frac{d}{D}$$

d = distance from original spotting line to the *center* of the final spot.
D = distance from original spotting line to the solvent front.

Calculate the R_f values for each cation spot in the unknown. Compare these with the known cation R_f values.

Record the R_f values under DATA.

By observing the colors, the positions, and R_f values of the unknown cation spots compared to those of the known cation spots, determine which cations you have in your unknown. In the report CONCLUSION section, circle those that are present in your unknown and summarize how you arrived at your conclusion.

Name   **Instructor/Section**   **Lab Partner**   **Date**

Experiment 6: Chromatography

DATA

	Colors					
	Mn^{2+}	Fe^{3+}	Co^{2+}	Ni^{2+}	Cu^{2+}	Unknown
The original spot colors	_____	_____	_____	_____	_____	_____
After developing and drying	_____	_____	_____	_____	_____	_____
After ammonia, NH_3	_____	_____	_____	_____	_____	_____
After sodium hydroxide, NaOH	_____					_____
After dimethylglyoxime				_____		_____
R_f value =	_____	_____	_____	_____	_____	_____

CONCLUSION

Unknown number _____

Circle cations present Mn^{2+} Fe^{3+} Co^{2+} Ni^{2+} Cu^{2+}
in the unknown

Discussion (what were you able to accomplish):

Experiment 6: PRE-LAB or POST-LAB QUIZ

Multiple choice questions. Circle your answers.

1. The type of chromatography used in this experiment is

   a. gas chromatography.
   b. thin layer chromatography.
   c. column chromatography.
   d. paper chromatography.

2. The distance d (small letter) represents the distance from the original spotting line to the

   a. top of the final spot.
   b. center of final spot.
   c. bottom of the final spot.
   d. solvent front.
   e. bottom edge of the chromatogram.

3. The distance D (capital letter) represents the distance from the original spotting line to the

   a. top of the final spot.
   b. center of final spot.
   c. bottom of the final spot.
   d. solvent front.
   e. bottom edge of the chromatogram.

4. What is the R_f value for a spot where $D = 8.2$ cm and $d = 3.2$ cm?

MOLAR MASSES OF SOME GASES

7

PURPOSE

To determine the molar masses of several gases, based on oxygen as a standard.

PERFORMANCE OBJECTIVE

After performing these investigations, you should be able to experimentally determine the molar mass of an unknown gas. You should also be familiar with the use of ratios for comparing molar masses to a standard.

MATERIALS

Analytical balance, 500 mL or 250 mL plastic bottle with screw cap (bottle mass must be <100 g if weighed on an analytical balance), 200 or 250 mL graduated cylinder, various compressed gases including oxygen, argon, propane, natural gas, carbon dioxide, and others.

PROCEDURE

Weigh a clean, dry 500 mL plastic bottle with cap to the nearest 0.0001 g and record its mass. Be sure to use the same analytical balance for each weighing, and that it is zeroed properly. Flush the bottle with the gas to be used. Put the tubing from the tank of gas (set at 10 psi) all the way into the bottle while counting slowly to ten or twelve. Cap the bottle quickly to avoid mixing the gas with air from the room. Invert the bottle while collecting natural gas, helium, hydrogen, and other gases that are less dense than air. Do not squeeze the bottle; hold it only by the bottle cap. Record the mass of the bottle with each of the gases being investigated.

Finally, fill the bottle to the top with tap water, and pour the water into a graduated cylinder and record the volume of the bottle. Calculate the mass of air in the bottle by using the volume of the bottle and the density of air from the table provided.

Density of Dry Air in g/mL
Source: CRC Handbook of Physics and Chemistry

°C \ P	Centimeters of Mercury			
	73.0	74.0	75.0	76.0
20	0.001157	0.001173	0.001189	0.001205
21	0.001153	0.001169	0.001185	0.001201
22	0.001149	0.001165	0.001181	0.001197
23	0.001145	0.001161	0.001177	0.001193
24	0.001142	0.001157	0.001173	0.001189
25	0.001138	0.001153	0.001169	0.001185
26	0.001134	0.001149	0.001165	0.001181

Subtract the calculated mass of air in the bottle from the mass of the bottle plus air to obtain the mass of the empty bottle (line e below table). Enter this value in line 3. The mass of each gas sample (line 4) is obtained by subtracting line 3 from line 2.

Divide the mass of each gas by the mass of oxygen present in the bottle (given in line 4) to obtain the relative masses of the gases compared to oxygen, and record these values in line 5. Multiply each mass in line 5 by 32 to obtain relative masses of the gases based on oxygen having a mass of 32.0 g/mol.

Use molar masses from the periodic table to determine accepted (true) molar masses and record these in line 7. Finally, determine the difference between experimental and accepted values (line 8) and the percent error (line 9). The percent error is the difference between experimental and accepted values (line 8) divided by the accepted molar mass (line 7).

Errors of less than one percent are generally obtained by students for these investigations.

Experiment 7: Molar Masses of Gases

	Air	Oxygen	Argon	CO$_2$	Propane or other	Natural gas (the unknown)
1. Molecular Formula						unknown mixture
2. Mass of bottle with gas (to nearest 0.0001g)						
3. Mass of empty bottle* (calculated below)						
4. Mass of gas (line 2 − line 3)						
5. Mass of each gas (line 4) / Mass of O$_2$ from line 4						
6. EXPERIMENTAL MOLAR MASS of each gas (32.0 × previous line)						
7. ACCEPTED Molar Mass (from Periodic Table)**						
8. Difference between experimental & accepted Molar Masses						
9. Percent error = Difference / Accepted molar mass × 100%						

*MASS OF EMPTY BOTTLE:

    a. Mass of bottle + air (Line 2)   _____ g

    b. Volume of bottle   _____ mL

    c. Density of air in lab (from table)   _____ g/mL

    d. Mass of air in bottle = (vol. of bottle) × (density of air)   _____ g

    e. Mass of empty bottle = line a − line d   _____ g

** The accepted molar mass of air can be calculated by assuming it is 80% nitrogen, N$_2$, and 20% oxygen, O$_2$.

$$\text{Percent} \times \text{Molar Mass}$$
Nitrogen: (0.80) × (28) = 22.4
Oxygen: (0.20) × (32) = 6.4
Calculated Molar Mass of Air = 28.8 g/mol

Experiment 7: PRE-LAB or POST-LAB QUIZ

1. Circle names of gases that should be collected when the bottle is held upside down.

   Oxygen   Propane   Natural gas   Carbon dioxide   Helium

2. After weighing the gases, fill the bottle with tap water

   a. to overflowing and measure this volume with a graduated cylinder.
   b. to overflowing and weigh it.
   c. to the neck of the bottle and measure this volume with a graduated cylinder.
   d. to the neck of the bottle and weigh it.

3. In this experiment, molar masses of gases are based on

   a. air as a standard.
   b. oxygen gas as a standard.
   c. nitrogen gas as a standard.
   d. natural gas as a standard.
   e. carbon-12 as a standard.

ATOMS AND SPECTRA

PURPOSE

To study ways that we can use to identify elements by their spectra and to study methods of exciting electrons in atoms.

PERFORMANCE OBJECTIVES

Following this investigation, you should be personally convinced that electrons in atoms can be excited by light, heat, and electron bombardment (high voltage), that atoms differ from one another, and that we can identify them by spectroscopic methods.

BACKGROUND EXPLANATION

An atom of helium has a diameter of about 1×10^{-8} cm or 100 picometers. That is far too small to be detected by the naked eye. In May, 1970, *Chemical & Engineering News* (*C&EN*) reported that a University of Chicago physicist, Albert Crewe, had "seen" single uranium and thorium atoms magnified a million times by a scanning electron microscope. The uranium atom has a radius of about 140 pm. The investigations we will perform allow us to identify various kinds of atoms by their spectra.

MATERIALS

Spectra tubes, spectrum tube power supply, Crookes tube, magnet, simple grating spectroscopes, samples of various compounds in solution, nichrome wire in handles for flame tests, ultraviolet lamp, an incandescent lamp, a heat lamp, and a hot plate.

PROCEDURE

Many of these investigations may be done as a group under the direction of your lab instructor.

A. Crookes tube experiment: Apply a high voltage source to one terminal while grounding the other terminal of a Crookes tube. What do you observe? Slowly move a

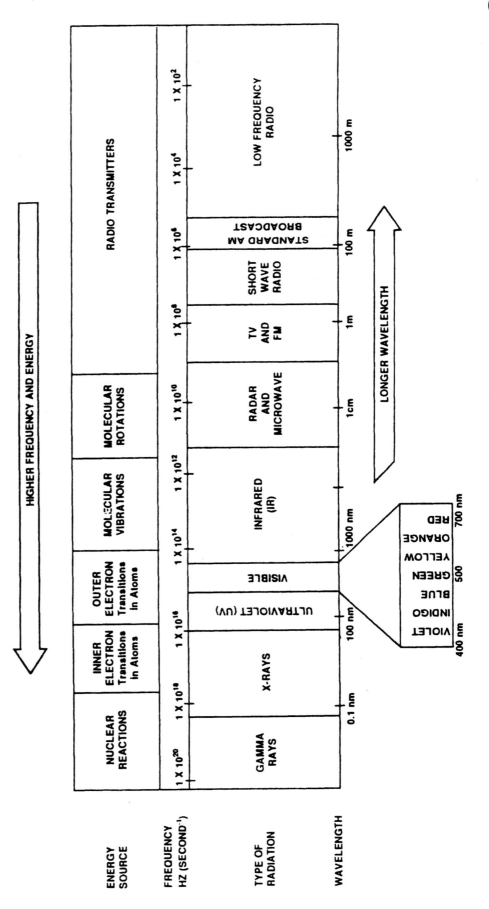

THE ELECTROMAGNETIC SPECTRUM

horseshoe magnet near the top of the tube, and perpendicular to it. Reverse the polarity of the magnet by turning it around. What do you observe?

B. Gas spectra: Connect gas discharge tubes containing various gases to the spectrum tube power supply, observe the line spectra through the grating spectroscope, and record your observations.

C. Cation spectra: Use a nichrome wire with a loop in one end and a cork for a handle at the opposite end. Place the loop end of the wire in a sample of a salt solution. Place the wire quickly in the burner flame and observe the color with the naked eye and again view it with a grating spectroscope. Solutions tested should include those listed in the report sheet.

D. Study the figure showing regions of the Electromagnetic Spectrum to see the relation of wavelength, frequency, energy, and energy sources for the various forms of electromagnetic radiation.

E. Turn on an ultraviolet light (black light), an incandescent bulb with tungsten filament, a heat lamp, and a hot plate (infrared). Bring your hand near each one to sense heat, and record your observations. Turn off the lights in the room and bring a piece of white cloth near each one, and record your observations.

Experiment 8: Atoms and Spectra

Part A. Crookes tube experiment:

1. Crookes tube observed initially using high voltage: _____

2. Effect of horseshoe magnet: _____

3. Effect of reversing the magnet: _____

4. Explanation of the above phenomenon: _____

_____

Part B. Gas spectra:

	Gas Observed	Color to naked eye	Characteristic colors of lines viewed with spectroscope
1	Mercury		
2	Helium		
3	Neon		
4	Argon		
5	Krypton		
6	Oxygen		
7	Hydrogen		
8			

Part C. Cation Spectra:

	Compound tested	Ion responsible for color	Color to naked eye	Characteristic line colors viewed with spectroscope
1	Lithium chloride			
2	Sodium chloride			
3	Potassium chloride			
4	Strontium chloride			
5	Calcium chloride			
6	Copper(II) chloride			
7	Unknown			

Part E.

Energy source	Qualitative heat measure	Observed effects of light source on white cloth in darkened room.
Ultraviolet lamp		
Incandescent lamp		
Heat lamp		
Hot Plate (IR)		

ATOMS AND SPECTRA QUESTIONS

1. What is line spectra and explain why it is important to chemical analysis?

2. What do the distinct lines in a line spectra reveal about atoms? Why does each element exhibit a line spectra instead of a continuous spectrum?

3. What form of electromagnetic radiation has energy slightly higher than that of visible light?

4. What form of electromagnetic radiation has energy slightly lower than that of visible light?

5. Given that the visible spectrum includes the colors red, orange, yellow, green, blue, indigo, and violet (ROY G BIV is an acronym), indicate which one of these colors has

   the longest wavelength _____   the shortest wavelength _____

   the highest frequency _____   the greatest energy _____

6. Why can you see certain objects under a "black light" (ultraviolet light)?

7. Heat is infrared radiation, so why can you see objects placed under a "heat lamp" which emits infrared radiation?

 ______________________ ______________________
Name Instructor/Section

Experiment 8: PRE-LAB or POST-LAB QUIZ

Multiple choice questions. Circle your answers.

1. Pictures of individual atoms
   a. were first seen in 1930.
   b. were first seen in 1950.
   c. were first seen in 1970.
   d. were first seen in 1990.
   e. have not yet been obtained.

2. Which color of visible light listed here has the longest wavelength?
   a. violet
   b. blue
   c. yellow
   d. red
   e. green

3. Which form of electromagnetic radiation listed here has the longest wavelengths?
   a. visible
   b. UV
   c. X-rays
   d. gamma rays
   e. IR

4. Which form of electromagnetic radiation listed here is sometimes called "black light"?
   a. UV
   b. IR
   c. visible
   d. X-rays
   e. gamma rays

5. Give three ways by which electrons in atoms can be excited. ______________________

____

CONDUCTIVITY AND SOLUBILITY PROPERTIES

9

(Performed live by lab instructor or as a video)

PURPOSE

To study the conductivity characteristics and solubilities of several chemical substances in relation to bonding characteristics.

MATERIALS

Chemicals listed in the data table, conductivity apparatus consisting of l00w, 15w, and 0.5 watt neon glower bulb, each wired in parallel with one another and in series with electrodes.

PROCEDURE

Test the materials listed for conductivity and solubility characteristics, and give the bonding for each.

Experiment 9: Conductivity and Solubility Properties

CONDUCTIVITY AND BONDING DATA

Material	Conductivity (check the appropriate category)			BONDING choices: ionic, covalent, or intermediate
	Good	Poor	Very poor or nonconductor	
1. KNO_3 (s)				KNO_3 is
2 KNO_3 (aqueous)				
3. KNO_3 (fused)				
4. NaCl (s)				NaCl is
5. NaCl (aqueous)				
6. Table sugar (s)				Sugar is
7 Sugar (aqueous)				
8. Sugar (fused)				
9. Ethanol, C_2H_5OH				
10. Benzene, C_6H_6				
11. H_2O (distilled)				
12. Tap water				
13. Toluene, $C_6H_5-CH_3$				
14. CCl_4				
15. H_2O + C_6H_6 (benzene)				
16. HCl(g) bubbled in H_2O layer				HCl(aq) is
17. HCl(g) bubbled in C_6H_6 layer				HCl(g) is
18. Acetic acid (conc.) CH_3COOH				
19. Acetic acid (0.1M) CH_3COOH				

SOLUBILITY DATA AND BONDING

Mixture	Miscible	Immiscible	Bonding in second member of pair (check appropriate category)		
			Polar	Intermediate	Nonpolar
20. H_2O + benzene					
21. H_2O + ether					
22. H_2O + CCl_4					
23. CCl_4 + acetone					
24. H_2O + acetone					

Experiment 9: POST-LAB QUIZ

Multiple choice questions. Circle your answers.

1. Based on conductivity data, the bonding in potassium nitrate, KNO_3, is

   a. ionic.
   b. covalent.
   c. intermediate (between ionic and covalent).
   d. uncertain—the data is too ambiguous to make a decision.

2. Carbon tetrachloride, CCl_4, is

   a. a good conductor of electricity.
   b. a poor conductor of electricity.
   c. a nonconductor of electricity.

3. Based on conductivity data, the bonding in carbon tetrachloride, CCl_4, is

   a. ionic.
   b. covalent.
   c. intermediate (between ionic and covalent).
   d. uncertain—the data is too ambiguous to make a decision.

4. Hydrogen chloride gas, HC1, in benzene is a _____(1)_____ .

   Hydrogen chloride gas, HC1, in water is a _____(2)_____ .

   Circle ONE letter with the best answers for both statements.
   a. (1) GOOD conductor of electricity.   (2) GOOD conductor of electricity.
   b. (1) POOR conductor of electricity.   (2) POOR conductor of electricity.
   c. (1) NONconductor of electricity.   (2) NONconductor of electricity.
   d. (1) GOOD conductor of electricity.   (2) NONconductor of electricity.
   e. (1) NONconductor of electricity.   (2) GOOD conductor of electricity.

5. Water and carbon tetrachloride are _____(1)_____ .

   Carbon tetrachloride molecules are _____(2)_____ .

   Circle ONE letter with the best answers for both statements.
   a. (1) miscible   (2) polar
   b. (1) miscible   (2) nonpolar
   c. (1) immiscible   (2) polar
   d. (1) immiscible   (2) nonpolar

MOLECULAR MODELS AND SHAPES OF MOLECULES

10

PURPOSE

To study the bonding and shapes of molecules.

PERFORMANCE OBJECTIVES

Following this investigation you should be familiar with the structures and shapes of representative molecules.

BACKGROUND EXPLANATION

Shapes of molecules are quite important in understanding the chemistry of molecules. Water, for instance, is known to have a "bent" or angular shape, as shown below, with a bond angle of 105°. Its chemistry is very different from that of CO_2, which has a linear structure (all atoms in a line).

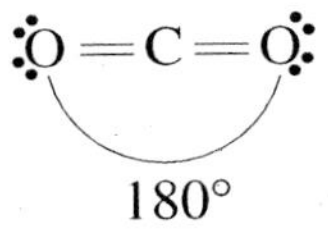

Note that a linear molecule has a 180° bond angle. Much of the difference in the chemistry of these molecules originates from the difference in their shapes. Therefore, it is important that we understand the shapes of molecules. Molecular models are useful tools for helping us understand these shapes.

The most common type of molecular model is made from balls and sticks, or a combination of sticks and springs and balls. The balls represent atoms and the sticks or springs represent bonds. Ball-and-stick models are used to represent molecules that are covalently bonded so each molecule has a definite number of atoms and a definite number of covalent bonds.

In the model kit, balls with different colors represent different atoms. For each kind of atom, the ball has the correct number of holes drilled at the proper angles so the model will give a representation of the correct shape of the molecule. Thus, by building the model, we can learn a lot about the shape of the molecule. In many model kits, hydrogen atoms are represented by yellow balls (1 hole), oxygen atoms by red balls (2 holes), carbon atoms by black balls

(4 holes), and nitrogen atoms by blue balls. The halogens, (chlorine, bromine, and iodine) are represented by green, orange, and violet colored balls, respectively, each with one hole. Molecules come in a variety of shapes, including

Linear: All atoms are in a straight line.

Bent (angular): Molecules consisting of three atoms not arranged in a straight line.

Trigonal Planar: Has three atoms arranged at the corners of a triangle around a central atom, all in the same plane with bond angles of 120° at the central atom.

Trigonal Pyramidal: Looks like a tripod with three atoms joined to a fourth atom. It also looks like a tetrahedron that is missing one atom.

Tetrahedral: Has four atoms bonded to one central atom, all at angles of 109.5°.

Planar: Has four or more atoms all in a plane.

MOLECULAR STRUCTURES: THE VSEPR MODEL

Another way to discover the shape of a molecule is to use a theory called the VSEPR (valence shell electron pair repulsion) theory. The main idea of this theory is that pairs of electrons around the central atom repel each other so they stay as far apart as possible. Let us see how this works. If we look at the molecule $BeCl_2$ and write its Lewis structure

$$Cl\text{---}Be\text{---}Cl$$

we will see that there are only two pairs of electrons around the beryllium atom. What arrangement will allow these two pairs to be as far apart as possible? The best arrangement is for the shared pairs to be on opposite sides of the beryllium atom, 180° from each other. That is the maximum possible separation; it gives the molecule a linear shape. Whenever two pairs of electrons are present around an atom, they should always be placed at an angle of 180° to each other to give a linear arrangement.

Let's look at BF_3. If we draw its Lewis structure we will find that boron is surrounded by three pairs of electrons.

$$F \diagdown \quad \diagup F$$
$$B$$
$$\mid \diagup 120°$$
$$F$$

The arrangement that minimizes the repulsions and puts the pairs as far apart as possible has all three pairs separated by an angle of 120° and with all atoms in the same plane. This flat planar molecule with a triangular arrangement of atoms around the central atom is known as trigonal planar. Whenever three pairs of electrons are present around an atom, the atoms should always be placed at the corners of a triangle in a plane at an angle of 120° to each other.

Now let's look at a molecule that has four pairs of electrons around the central atom. If we draw the Lewis structure of the molecule of CH_4 we will see that there are four pairs of electrons around the central carbon atom. The best way to get the largest distance between the four pairs is to have the bond angles at approximately 109.5°. The carbon sits in the middle of a tetrahedron (a tetrahedron has four equal, triangular faces). Whenever four pairs of electrons are present around an atom, they should always be placed at the corners of a tetrahedron.

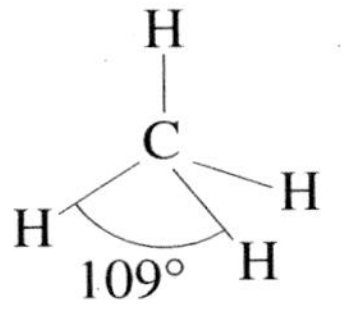

One other shape needs to be discussed. If we look at the Lewis structure of NH_3, we find that there are four pairs of electrons around the central nitrogen atom. These four pairs of electrons are in a tetrahedral arrangement around the nitrogen. However, only three pairs of the electrons are bonded to hydrogen atoms; one pair is an unshared pair. Although the arrangement of electron pairs is tetrahedral, the hydrogen *atoms* occupy only three of the four corners, so the molecule has a trigonal pyramidal structure.

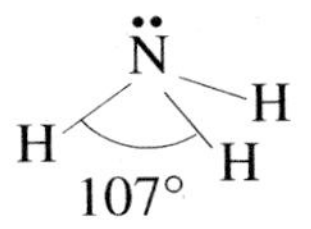

Lastly, we will need to work with molecules having double or triple bonds. When using VSEPR theory, and when the Lewis structure shows that there is a double bond such as in formaldehyde shown below, the two pairs of electrons in the double bond are counted as one (after all, the two pairs are held in the same location because they are in the same bond). Thus, for the molecule below, the central carbon atom acts like it has only three electron pairs, instead of four, so the shape of the molecule is trigonal planar with bond angles of 120°.

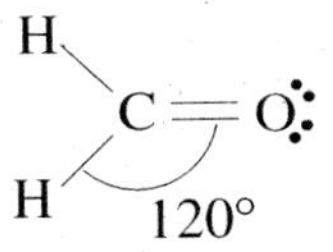

PROCEDURE

Completion of the experiment will involve determining the number of valence electrons in a molecule, writing the correct Lewis structure, and then constructing the model to see if the shape predicted by VSEPR is the shape formed by the models.

In determining the number of electrons in the Lewis structure, remember to count only the outer valence electrons. For example, suppose we want to determine the total number of valence electrons in ethyl alcohol, C_2H_5OH. Carbon is in Group 4 of the periodic table and has only 4 valence electrons. Hydrogen, in Group 1, has only one valence electron, and oxygen, in Group 6, has six valence electrons. Therefore,

$$
\begin{aligned}
2\,C \times 4\,e^- &= 8 \text{ valence electrons} \\
6\,H \times 1\,e^- &= 6 \text{ valence electrons} \\
1\,O \times 6\,e^- &= 6 \text{ valence electrons} \\
\hline
& 20 \text{ valence electrons}
\end{aligned}
$$

A correct Lewis structure will have twenty electrons. Using the rules for Lewis structures, we can write a structural formula, as shown below, using dashes to represent a pair of shared electrons, and pairs of dots for unshared pairs of electrons.

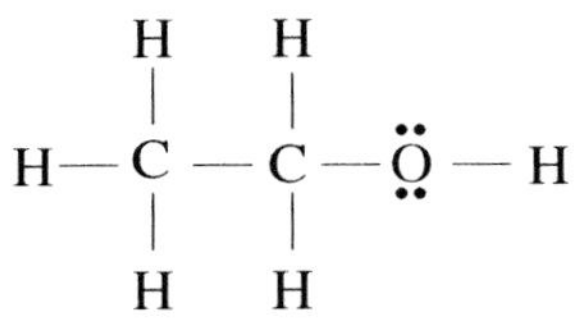

$$
\begin{array}{ccccc}
 & H & & H & \\
 & | & & | & \\
H - & C & - & C - \ddot{\underset{\cdot\cdot}{O}} & - H \\
 & | & & | & \\
 & H & & H & \\
\end{array}
$$

For each structure, draw the Lewis structure, construct the model, record the observed shape, and describe polarity.

Experiment 10: Shapes of Molecules

Name	Chemical formula	Combined Dot-dash formula	Shape	Polar or Nonpolar
Water	H_2O			
Hydrogen bromide	HBr			
Ammonia	NH_3			
Methane (the main chemical in natural gas)	CH_4			
Ethane	C_2H_6			
Ethene or Ethylene (used to make polyethylene plastic)	C_2H_4			
Ethyne or Acetylene (used in welding)	C_2H_2			
Hydrogen cyanide (a lethal gas)	HCN			
Ethanol (ethyl alcohol)	C_2H_5OH			
Carbon tetrachloride	CCl_4			
Carbon dioxide	CO_2			
Hydrogen sulfide (rotten-egg gas)	H_2S			
Nitrogen triiodide	NI_3			
Hydrogen peroxide (used as a bleach)	H_2O_2			

Without building models, predict the shape of the following molecules by first calculating the number of valence electrons, writing the Lewis structure, and then using VSEPR theory to predict the shape of the molecule. Draw a structural formula to represent each one.

1. Beryllium difluoride BeF_2
   (First locate Be on the periodic table
   and note its number of valence electrons.
   It does not obey the octet rule in bonding.)

2. Phosphorous trichloride PCl_3
   (First locate P on the periodic table
   and note its number of valence electrons.)

3. Aluminum trichloride $AlCl_3$
   (Again, use the periodic table.)

Experiment 10: PRE-LAB or POST-LAB QUIZ

Multiple choice questions. Circle your answers.

1. A water molecule has a _____ shape.

    a. linear   Draw its structure here:
    b. bent (angular)
    c. trigonal planar
    d. trigonal pyramidal
    e. tetrahedral

2. All atoms in a CO_2 molecule are in a straight line so CO_2 has a _____ shape.

    a. linear   Draw its structure here:
    b. bent (angular)
    c. trigonal planar
    d. trigonal pyramidal
    e. tetrahedral

3. Molecules of methane, CH_4, and carbon tetrachloride, CCl_4, have bond angles of 109.5 degrees, so they have a _____ shape.

    a. linear   Draw the CH_4 structure here:
    b. bent (angular)
    c. trigonal planar
    d. trigonal pyramidal
    e. tetrahedral

4. Ammonia, NH_3, molecules have one lone pair of electrons and three nitrogen-to-hydrogen bonds, so ammonia has a _____ shape.

    a. linear   Draw its structure here:
    b. bent (angular)
    c. trigonal planar
    d. trigonal pyramidal
    e. tetrahedral

THE FORMULA OF A HYDRATE

11

PURPOSE

To determine the formula for a hydrated compound and to determine hygroscopic or efflorescent properties of some substances.

PERFORMANCE OBJECTIVES

Following this investigation you should be able to determine the percentage of water in a hydrate, and also determine its formula. You should be able to do calculations similar to those encountered while carrying out this investigation. You should also be familiar with the new terms and techniques described.

BACKGROUND EXPLANATION

Paper, wood, and many other chemical substances that have been exposed to the atmosphere contain water. This water can usually be removed by gently heating; it is not present in the material in any definite amount. For other substances, generally solid ionic compounds, larger amounts of water are rather strongly bound to the compound, and in definite proportions with the compound. This water of hydration is generally bound to the cations as a part of the crystal structure, but heating gives the anhydrous salt and water. We can represent the process by the following equation:

$$Ba(OH)_2 \cdot 8H_2O \longrightarrow Ba(OH)_2 + 8H_2O$$

$$\text{Hydrate} \qquad\qquad \text{Anhydrous} \quad \text{Water}$$
$$\text{salt}$$

Some hydrated compounds lose water spontaneously to the atmosphere when left in the open. These compounds are EFFLORESCENT. Other compounds are anhydrous (without water) and will absorb water from the atmosphere. These compounds are HYGROSCOPIC. When absorption is great enough for the substance to be used as a drying agent for gases or liquids, it is called a DESICCANT. A few compounds will take up so much water that they eventually dissolve in their water of hydration. These substances are DELIQUESCENT. In this experiment you will investigate the efflorescent and hygroscopic nature of some compounds, and determine the formula for an unknown hydrate.

PROCEDURE

Note: To utilize your time most efficiently it is suggested that you begin part C first, then start part B, and finally do part A.

A. Which compounds are hydrates? Test each of the following compounds for water of hydration by placing a small amount (about 0.5 g, the size of a pea) in a small dry test tube, and by heating GENTLY for one minute. If droplets of water condense on the cool upper walls while the bottom of the test tube is heated gently, indicate on your report sheet that water was produced during heating. Compare the color of the residue to the original sample. Test 2 small crystals of the residue to see if it is water soluble by trying to dissolve them in a 250 mL beaker half full of water. If water was given off, but the residue is insoluble, then the compound is one that decomposes when heated, like sugar, but such compounds are not hydrates. To be a hydrate, the compound must give water upon heating, and must be water soluble.

Test these compounds as described:

Barium chloride, Sodium chloride, Copper sulfate (blue crystals)

B. What are some compounds that are efflorescent or hygroscopic? Place a few crystals of each of the following compounds on separate weighing boats or squares of paper with edges folded like a box. Rapidly weigh each sample and container to the greatest accuracy, depending on the balance used. Then, reweigh each container and sample one hour or more later. Record your results on the data table provided. Test the following chemicals:
$CaCl_2$ (anhydrous)
NaOH pellets   **CAUTION: Do not touch NaOH; it is caustic.**
$ZnSO_4 \cdot 7H_2O$

C. Analysis of water of hydration: Heat a clean crucible and cover held by a clay triangle over a blue flame to remove any moisture in the porcelain. For the hottest flame,

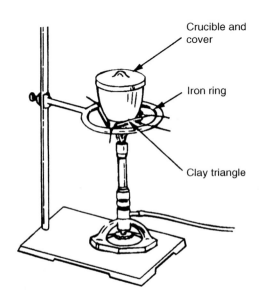

Crucible and cover

Iron ring

Clay triangle

adjust the burner so the tip of the blue inner cone of the flame is about 1.0 cm below the bottom of the crucible.

Heat the crucible to redness, and then allow it to cool to room temperature on a ceramic square using tongs—not your fingers—to handle the crucible and lid. Weigh the crucible and lid on an analytical balance.

Fill the crucible one-half full with your unknown solid hydrate, and reweigh the crucible and lid.

Support the crucible on the clay triangle with the crucible lid in place during the first five minutes of gentle heating.

Heat the crucible with the lid tipped and touching a corner of the triangle while heating to redness for another ten to fifteen minutes. Put the lid back in place on the crucible.

Allow the crucible to cool to room temperature; then weigh the crucible, lid and residue. Record this mass.

Repeat the heating and reweigh the crucible to be sure the compound is entirely dehydrated. Calculate the formula for the hydrate.

Report Sheet ______________________________ ______________ ______________ ______________
Name Instructor/Section Lab Partner Date

Experiment 11: Hydrates

Water of Hydration Data:

Part A. Which compounds are hydrates?

Compound	Color before heating	Color after heating	Did water appear?	Was the residue soluble?	Is the compound a hydrate?
Barium chloride					
Sodium chloride					
Copper sulfate					

Part B. Which compounds are efflorescent, hygroscopic, or neither?

Compound	Total Initial Weight	Total Final Weight	CONCLUSION: Efflorescent Hygroscopic No change
$CaCl_2$ (anhydrous)			
NaOH (pellets)			
$ZnSO_4 \cdot 7H_2O$			

Part C. Water of hydration analysis

1. Formula of anhydrous salt used
   (formula of the salt without water)   _____

2. Mass of dry crucible and cover after preheating   _____ g

3. Mass of crucible, cover, and the hydrate   _____ g

4. Mass of crucible, cover, and residue remaining
   after heating 15-20 min.   _____ g

5. Mass of crucible, cover, and residue remaining
   after additional heating   _____ g

6. Mass of hydrate (#3 minus #2)   _____ g

7. Mass of residue, the anhydrous salt
   (smaller of #4 or #5 minus #2)   _____ g

8. Mass of water lost (#6 minus #7)   _____ g

9. Molar mass (MM) of anhydrous salt   _____ g/mol

10. Molar mass (MM) of water   _____ g/mol

11. Moles of anhydrous salt $= \dfrac{\text{mass of anhydrous salt, \#7}}{\text{MM anhydrous salt, \#9}}$ _____ mol

12. Moles of water $= \dfrac{\text{mass of water lost, \#8}}{\text{MM of water, \#10}}$ _____ mol

13. Moles of water per mole of anhydrous salt (#12/#11)
    (Give answer to three significant figures.)   _____

14. FORMULA OF HYDRATE when water is rounded to
    nearest multiple of one-half. (Example: $CuSO_4 \cdot 5\ H_2O$)   _____

Experiment 11: PRE-LAB or POST-LAB QUIZ

Multiple choice questions. Circle your answers.

1. A substance that loses water when heated and also gives a residue that is water soluble is

    a. hygroscopic.
    b. a hydrate.
    c. a desiccant.
    d. deliquescent.
    e. efflorescent.

2. The solid residue that remains when a hydrate is heated is

    a. an anhydrous salt.
    b. a hygroscopic compound.
    c. a desiccant.
    d. deliquescent.
    e. efflorescent.

3. The anhydrous compound that will absorb water from the atmosphere is

    a. a hydrate.
    b. hygroscopic.
    c. deliquescent.
    d. a desiccant.
    e. efflorescent.

4. A compound that will lose water spontaneously to the atmosphere is

    a. a hydrate.
    b. hygroscopic.
    c. deliquescent.
    d. a desiccant.
    e. efflorescent.

5. What would you observe when a pea-size amount of a hydrate is heated in a test tube?

IONIC REACTIONS

12

PURPOSE

To study what takes place during ionic reactions in solution, and to be able to write balanced chemical equations for these reactions.

PERFORMANCE OBJECTIVE

After completing these investigations you should be able to write chemical equations, ionic equations, and net ionic equations. You should also be able to predict products of reactions similar to the ones carried out.

BACKGROUND EXPLANATION

Many ionic compounds are soluble in water. When dissolved, their ions are dissociated (separated) and totally hydrated (surrounded by water molecules). Dissociated ions move randomly in the aqueous (water) solution. Some ionic compounds appear to be insoluble—or only slightly soluble—in water so they are classified as precipitates.

In this experiment we will be examining what happens when two aqueous solutions of soluble ionic compounds are mixed. When the solutions are mixed, the positive and negative ions are free to move about and to be attracted by ions with opposite charges. If attractions are great enough, an insoluble solid substance—called a **precipitate**—is formed. We can write the general equation for a double replacement reaction as

$$AB + CD \longrightarrow AD + CB$$

where AB exists as A^+ ions and B^- ions in solution and CD exists as C^+ ions and D^- ions in solution. As the ions move around, A^+ ions will come in contact with D^- ions and C^+ and B^- ions will come in contact so compounds AD and CB can be formed. If either one of these new compounds has a low solubility, a solid precipitate is formed. The formation of a precipitate is evidence that a chemical reaction has taken place.

This type of reaction is called a double replacement (or metathesis) reaction. When writing the chemical equation, the hypothetical reaction products are obtained by simply switching partners for the two reactants. (A positive ion can only have a partner that is a negative ion. The symbol of a positive ion is followed by the symbol of its partner, a negative ion.) The formula for each compound must be written correctly, with subscripts chosen so that each compound is neutral in charge. Once all chemical formulas are written *correctly*, the equation is balanced by using appropriate coefficients in front of formulas.

If a precipitate forms when two known solutions are mixed, solubility rules can be used to help us determine which product can be a precipitate. For this experiment, only four solubility rules are necessary.

SOLUBILITY RULES FOR THIS EXPERIMENT

Rule 1. All compounds containing nitrates are soluble. They dissolve in water.

Rule 2. All compounds containing Group IA metal ions are soluble. They dissolve in water.

Rule 3. All compounds containing the ammonium ion are soluble in water.

Rule 4. All reagent solutions provided for this experiment contain compounds that are soluble in water. (Use only when Rule 1, 2, and 3 do not apply.)

Let us look at two examples.

EXAMPLE 1: We can write a hypothetical equation for what might happen if we mix solutions of sodium chloride and potassium nitrate by simply switching ion partners to give

$$NaCl \ + \ KNO_3 \ \longrightarrow \ NaNO_3 \ + \ KCl$$

However, writing this equation does *not* mean that a reaction must occur. If we were to mix these two chemicals, we would not see a precipitate or any other evidence that a reaction has occurred. Here, both hypothetical products are water soluble according to solubility Rules 1 and 2. What we actually have in the solution is a mixture of the four kinds of ions, Na^+, Cl^-, K^+, and NO_3^-. The best written expression for this investigation is

$$NaCl \ + \ KNO_3 \ \longrightarrow \ No \ reaction$$

EXAMPLE 2: When solutions of sodium chloride and silver nitrate are mixed, the hypothetical equation for what might happen would be

$$NaCl \ + \ AgNO_3 \ \longrightarrow \ NaNO_3 \ + \ AgCl$$

When these two solutions are mixed, a white precipitate can be observed. This is evidence that a chemical reaction has occurred. According to solubility Rule 1, we know that all nitrates are soluble and, according to solubility Rule 2, all compounds containing Group IA metal ions are soluble, so the new compound, $NaNO_3$, must be soluble on both counts; it could not be the precipitate. Thus, the AgCl must be the precipitate, the insoluble product. This insoluble product is identified in the equation by underlining its formula and by placing the (s) symbol after its formula, as shown here.

$$NaCl \ + \ AgNO_3 \ \longrightarrow \ NaNO_3 \ + \ \underline{AgCl}(s)$$

This equation can also be written as an **ionic equation** to indicate which ions are dissolved and dissociated in water (aq) and which ions form insoluble, solid (s) compounds. The ionic equation for this reaction is shown here.

$$Na^+(aq) + Cl^-(aq) + Ag^+(aq) + NO_3^-(aq) \longrightarrow Na^+(aq) + NO_3^-(aq) + \underline{AgCl}(s)$$

The only reaction that actually occurs is between Ag^+ ions and Cl^- ions to form $AgCl(s)$. The sodium and nitrate ions remain in solution without reacting, so they are called **spectator ions**. Leaving out the spectator ions gives the **net ionic equation** shown here.

$$Ag^+(aq) + Cl^-(aq) \longrightarrow \underline{AgCl}(s)$$

PROCEDURE

WEAR GOGGLES TO PROTECT YOUR EYES

Mix approximately one milliliter of each of the two solutions specified in a test tube (or on a plastic well plate). Agitate the mixture and wait one to two minutes. The formation of a precipitate—a solid or cloudy product—is evidence that a chemical reaction has occurred.

Use the solubility rules to determine which one of the products is the precipitate. For each precipitation reaction, write a balanced equation, underline the formula of the precipitate, and also identify it with the "(s)" symbol. Specify the color of the precipitate and the solubility rule number(s) used to determine which product is the precipitate.

On the second blank line write the net ionic equation, showing only formulas of the ions involved in the reaction and the precipitate. If no reaction occurred, write the words "No reaction" in place of the reaction products.

Name   **Instructor/Section**   **Lab Partner**   **Date**

Experiment 12: Ionic Reactions

	Precipitate Color	Solubility Rule Number

Example:

Mix and briefly agitate 1 mL portions of
silver nitrate and potassium chromate   _____brown_____   _____Rule 1_____

    Line 1, chemical equation: $2\,AgNO_3 + K_2CrO_4 \longrightarrow \underline{Ag_2CrO_4(s) + 2\,KNO_3}$

    Line 2, net ionic equation: $2\,Ag^+ + CrO_4^{2-} \longrightarrow \underline{Ag_2CrO_4(s)}$

Part 1:

Mix and briefly agitate 1 mL portions of

    a. Lead(II) nitrate and potassium chromate   _____   _____

_____

_____

    b. Lead(II) nitrate and potassium dichromate   _____   _____

_____

_____

    c. Lead(II) nitrate and potassium iodide   _____   _____

_____

_____

    d. Barium chloride and potassium chromate   _____   _____

_____

_____

Part 2:

Mix and briefly agitate 1 mL portions of

    a. Sodium sulfate and barium chloride   _____   _____

_____

_____

    b. Copper(II) sulfate and barium nitrate   _____   _____

_____

_____

    c. Aluminum sulfate and barium nitrate   _____   _____

_____

Part 3:

Mix and briefly agitate 1 mL portions of

   a. Sodium chloride and silver nitrate   _____ _____

_____

_____

   b. Potassium bromide and silver nitrate   _____ _____

_____

_____

   c. Potassium bromide and ammonium chloride   _____ _____

_____

_____

   d. Sodium iodide and silver nitrate   _____ _____

_____

_____

Part 4:

Mix and briefly agitate 1 mL portions of

   a. Iron(III) chloride and barium hydroxide   _____ _____

_____

_____

   b. Cobalt(II) chloride and barium hydroxide   _____ _____

_____

_____

   c. Nickel(II) chloride and sodium hydroxide   _____ _____

_____

_____

   d. Copper(II) chloride and sodium hydroxide   _____ _____

_____

_____

 ______________________ ______________________
Name Instructor/Section

Experiment 12: PRE-LAB or POST-LAB QUIZ

Multiple choice questions. Circle your answers.

1 Precipitates form when compounds produced have

a. a very high solubility.
b. a moderately high or very high solubility.
c. a low solubility.

2. A positive ion can only combine with one that is negative.

a. true
b. false

3. Another name for metathesis reactions is

a. synthesis reactions.
b. oxidation-reduction reactions.
c. single replacement reactions.
d. double replacement reactions.
e. decomposition reactions.

4. When ionic compounds dissolve in water, the ions are "dissociated." Explain what "dissociated" means.

____

5. What compounds can be formed by the reaction of K_2SO_4 and $BaCl_2$?

____

6. Complete the following chemical equation:

$AgNO_3 + NaCl \rightarrow$ ______________________

7. Give the net ionic equation for the reaction indicated in question 6.

____

TYPES OF CHEMICAL REACTION
Introduction to Parts I-V

13

PURPOSE

To investigate a variety of chemical reactions and represent them by chemical equations.

PERFORMANCE OBJECTIVE

Following this exercise, you should be able to represent chemical reactions by writing balanced chemical equations. You should also be able to predict products of reactions if they are similar to those studied.

BACKGROUND EXPLANATION

The balanced chemical equation is used by chemists to summarize a chemical change. The writer of the equation first represents the reactants (starting materials) and products (substances formed) by writing a skeleton equation which includes the correct formulas for all substances involved.

In the equation for the combustion of propane

$$C_3H_8 + O_2 \longrightarrow CO_2 + H_2O$$

the reactants and products have been included, but the proper reaction proportions have not been indicated. Since atoms are conserved in every chemical reaction, coefficients are used to indicate the relative proportions of substances involved. The equation will then conform to the Law of Conservation of Mass.

Follow through with the steps used in balancing the equation shown here.

	Equation	Explanation
Step 1	$C_3H_8 + O_2 \longrightarrow 3CO_2 + H_2O$	1 mole of propane produces 3 moles of CO_2.
Step 2	$C_3H_8 + O_2 \longrightarrow 3CO_2 + 4H_2O$	1 mole of propane produces 4 moles of water.
Step 3	$C_3H_8 + 5O_2 \longrightarrow 3CO_2 + 4H_2O$	Balance oxygen atoms by using 5 moles of O_2.

PROCEDURE

The laboratory instructor will perform many of the following chemical reactions as demonstrations while you observe and record results. Many of these reactions require special precautions, and will therefore be performed for you, or videotapes of these demonstrations may be used. Feel free to ask the instructor any pertinent questions as you observe.

CHEMICAL REACTIONS — PART I: DECOMPOSITION

Certain compounds may be decomposed by heat, light, or electricity. In many instances, at least one of the reaction products for these reactions is a gas. Observe these chemical reactions as you complete and balance the following chemical equations:

1. $H_2O(l)$ $\xrightarrow[\text{D.C}]{\text{electrolysis}}$ _____ + O_2 (g)

2. $HgO(s)$ $\xrightarrow{\text{heat}}$ _____ + _____

3. $PbO_2(s)$ $\xrightarrow{\text{heat}}$ _____ + _____

4. $KClO_3(s)$ $\xrightarrow{\text{heat}}$ _____ + _____

5. $NaClO_3(s)$ $\xrightarrow{\text{heat}}$ _____ + _____

You may conclude that *all* chlorates decompose to yield oxygen gas plus the metal chloride.

6. $NaNO_3(s)$ $\xrightarrow{\text{heat}}$ _____ + _____

You may conclude that nitrates decompose to yield nitrites and oxygen gas.

7. H_2O_2 $\xrightarrow[\text{MnO}_2 \text{ or KI}]{\text{catalyst}}$ _____ + _____

CHEMICAL REACTIONS — PART II: METALS

Reactive metals combine with oxygen to yield oxides of metals. In certain cases active metal peroxides and superoxides are formed, but we shall not present these special reactions here. Complete and balance these equations with the assumption that metal oxides are formed.

1. ___ Na + ___ O_2 $\xrightarrow{\text{room temp.}}$ ___ Na_2O

2. ___ K + ___ O_2 $\xrightarrow{\text{room temp.}}$ _____

3. ___ Mg + ___ O_2 $\xrightarrow{\text{burn}}$ _____

4. ___ Fe + ___ O_2 $\xrightarrow{}$ _____

Soluble metal oxides dissolve in water to give basic solutions.

5. Na_2O + H_2O $\longrightarrow$ _____

6. K_2O + H_2O $\longrightarrow$ _____

7. MgO + H_2O $\longrightarrow$ _____

8. Fe_2O_3 + $3H_2O$ $\longrightarrow$ _____

Elements in group IA (alkali metals) and Group IIA (alkaline earths) of the periodic table react with water to produce hydrogen gas and metal hydroxides. Balance these equations.

9. ___Na + ___H_2O $\longrightarrow$ _____ + $H_2(g)$

10. ___Li + ___H_2O $\longrightarrow$ _____ + _____

11. ___K + ___H_2O $\longrightarrow$ _____ + _____

12. ___Ca + ___H_2O $\longrightarrow$ _____ + _____

13. ___Mg + ___H_2O(steam) $\longrightarrow$ _____ + _____

Metals can be oxidized by nonmetals other than oxygen. In the following oxidation-reduction reactions, chlorine and iodine are the oxidizing agents. The elemental halogen is reduced to the halide.

14. ___Cu + ___$Cl_2(g)$ $\longrightarrow$ _____

15. ___Sb + ___$Cl_2(g)$ $\longrightarrow$ _____

16. ___Al + ___$I_2(s)$ $\longrightarrow$ _____

CHEMICAL REACTIONS — PART III: NONMETALS

Nonmetals react with oxygen to produce nonmetal oxides. Balance these equations.

1. ___ S_8 + ___O_2 $\longrightarrow$ ___ SO_2

2. ___C + ___O_2 $\longrightarrow$ _____

3. ___ P_4 + ___O_2 $\longrightarrow$ ___ P_4O_{10}

Phosphorus can also be oxidized by chlorine and bromine to produce phosphorus halides.

4. ___ P_4 + ___$Cl_2(g)$ $\longrightarrow$ ___ PCl_5

5. ___ P_4 + ___$Br_2(1)$ $\longrightarrow$ ___ PBr_3

Nonmetal oxides react with water to produce acids.

6. ___SO_2 + ___H_2O ⟶ H_2SO_3 (sulfurous acid)

7. ___SO_3 + ___H_2O ⟶ _____ (sulfuric acid)

8. ___P_4O_{10} + ___H_2O ⟶ _____ (phosphoric acid)

9. ___CO_2 + ___H_2O ⟶ _____ (carbonic acid)

10. ___N_2O_5 + ___H_2O ⟶ _____ (nitric acid)

11. ___NO_2 + ___H_2O ⟶ _____ + NO

More active halogens react with compounds containing less active halogens. The elemental halogen reactant is the oxidizing agent and is reduced. The order of halogen reactivities follows: F_2 > Cl_2 > Br_2 > I_2.

12. ___HI + ___Cl_2 ⟶ _____ + _____

13. ___KBr + ___Cl_2 ⟶ _____ + _____

14. ___KCl + ___Br_2 ⟶ _____

Balance these additional equations. If no reaction is predicted, write "no reaction."

15. ___F_2 + ___$NaCl$ ⟶ _____ + _____

16. ___Br_2 + ___NaF ⟶ _____ + _____

17. ___I_2 + ___$NaCl$ ⟶ _____ + _____

18. ___Br_2 + ___AlI_3 ⟶ _____ + _____

CHEMICAL REACTIONS — PART IV: METAL REPLACEMENT

Hydrogen gas is liberated when metals with $E°$ values greater than hydrogen react with acids. Consult the table of $E°$ values (replacement series) following Part V.

1. ___$Zn(s)$ + ___$HCl(aq)$ ⟶ $ZnCl_2(aq)$ + _____

2. ___$Ca(s)$ + ___H_2SO_4 ⟶ _____ + _____

3. ___$Mg(s)$ + ___H_3PO_4 ⟶ _____ + _____

The reduced form of any element will react with and reduce the oxidized form of another element when the second element has a lower $E°$ value. Consult the table of $E°$ values as you balance these. Write "no reaction" where appropriate.

4. ___Zn + ___$Pb(NO_3)_2$ ⟶ _____ + _____

5. ___Pb + ___$Cu(NO_3)_2$ ⟶ _____ + _____

6. ___$ZnSO_4$ + ___Cu ⟶ _____ + _____

7. ___$AgNO_3$ + ___Cu ⟶ _____ + _____

CHEMICAL REACTIONS — PART V: MISCELLANEOUS

Neutralization reactions: Acid + Base ⟶ Salt + Water

1. ___HCl(aq) + ___NaOH ⟶ NaCl + H_2O

2. ___HCl(aq) + ___$Ca(OH)_2$ ⟶ _____ + _____

3. ___H_2CO_3 + ___$Mg(OH)_2$ ⟶ _____ + _____

In this next reaction, note that the salt is Na_2HPO_4. Since one "H" was not neutralized, the compound is often called sodium phosphate, dibasic, to indicate two moles of sodium are in each mole of the salt.

4. ___H_3PO_4 + ___2NaOH ⟶ Na_2HPO_4 + _____

Carbonates release carbon dioxide when they react with an acid, or when heated strongly. Balance these equations:

5. ___$CaCO_3$ + ___HCl ⟶ ___CO_2 + ___H_2O + ___$CaCl_2$

6. ___$MgCO_3$ + ___H_2SO_4 ⟶ _____ + _____ + _____

7. ___$CaCO_3$ $\xrightarrow{\text{heat}}$ ___CaO + ___CO_2
limestone   quicklime

Double replacement (metathesis) reactions occur when ionic compounds exchange ions to form at least one compound which is thermodynamically more stable than the reactants were. One of the products is often a precipitate.

8. ___KI + ___$AgNO_3$ ⟶ _____ + _____

9. ___$CoCl_2$ + ___KOH ⟶ _____ + _____

10. ___$FeCl_3$ + ___NaOH ⟶ _____ + _____

REPLACEMENT SERIES
Oxidation Half-Reactions of Some Metals

Reduced Form		Oxidized Form			Oxidation Potential $E°$ volts
Li	$\rightarrow$	Li^+	+	e^-	3.09
K	$\rightarrow$	K^+	+	e^-	2.92
Ca	$\rightarrow$	Ca^{+2}	+	$2e^-$	2.87
Na	$\rightarrow$	Na^+	+	e^-	2.71
Mg	$\rightarrow$	Mg^{+2}	+	$2e^-$	2.37
Al	$\rightarrow$	Al^{+3}	+	$3e^-$	1.66
Zn	$\rightarrow$	Zn^{+2}	+	$2e^-$	0.76
Fe	$\rightarrow$	Fe^{+2}	+	$2e^-$	0.44
Cd	$\rightarrow$	Cd^{+2}	+	$2e^-$	0.40
Ni	$\rightarrow$	Ni^{+2}	+	$2e^-$	0.25
Sn	$\rightarrow$	Sn^{+2}	+	$2e^-$	0.14
Pb	$\rightarrow$	Pb^{+2}	+	$2e^-$	0.13
H_2	$\rightarrow$	$2H^+$	+	$2e^-$	0.00
Cu^+	$\rightarrow$	Cu^{+2}	+	e^-	–0.15
Cu	$\rightarrow$	Cu^{+2}	+	$2e^-$	–0.34
Cu	$\rightarrow$	Cu^+	+	e^-	–0.50
Fe^{+2}	$\rightarrow$	Fe^{+3}	+	e^-	–0.77
$2Hg$	$\rightarrow$	$2Hg_2^{+2}$	+	$2e^-$	–0.79
Ag	$\rightarrow$	Ag^+	+	e^-	–0.80
Au	$\rightarrow$	Au^{+3}	+	$3e^-$	–1.50

THE PREPARATION OF ALUM FROM SCRAP ALUMINUM

14

PURPOSE

To synthesize the compound alum from scrap aluminum.

PERFORMANCE OBJECTIVE

Following this experiment you should be able to calculate percentage yield and have an idea of how chemical processes can be used to transform scrap materials into useful substances.

MATERIALS

aluminum foil	melting point tube
250-mL beaker	rubber band
Buchner funnel	bottle for sample
filter paper	labels
short-stem funnel	potassium hydroxide
glass wool or cotton	solution (80 g/liter, 1.5M)
graduated cylinder	distilled water
laboratory burner	9M sulfuric acid
vacuum filter flask	ethyl alcohol
mortar and pestle	thermometer

BACKGROUND EXPLANATION

Some discarded materials can be transformed by chemical reactions into useful materials. The new materials formed have completely different physical and chemical properties from those of the original substances. Aluminum is used widely for beverage cans, but aluminum cans are more resistant to disintegration than are steel cans. This has led to growing concern regarding the disposal of aluminum containers.

This laboratory investigation presents a method for synthesizing one useful aluminum compound, alum, from scrap aluminum. Aluminum dissolves in a hot aqueous solution of potassium hydroxide. The equation for this reaction is:

$$1. \quad 2Al + 2KOH + 6H_2O \longrightarrow 2KAl(OH)_4 + 3H_2(g)$$

(in solution)

As this reaction proceeds, the concentration of the hydroxide ion in solution decreases. After the aluminum has been dissolved, it can be transformed into several compounds which may be useful as prepared, or useful as intermediates for the preparation of still other aluminum compounds.

Alums are a group of compounds that contain an alkali metal ion such as potassium or ammonium, and a tripositive ion such as Al^{+3} or Cr^{+3}, along with a negative ion, usually sulfate. Alums are hydrated with 12 water molecules. The alum prepared in this experiment is potassium aluminum sulfate, $KAl(SO_4)_2 \cdot 12H_2O$. It is used in medicine, in the dyeing of fabrics, in the manufacture of paper and pickles, for water purification, and for other industrial uses. Other alums include $AlNH_4(SO_4)_2 \cdot 12H_2O$, used as an astringent, and $CrK(SO_4)_2 \cdot 12H_2O$ used in photography.

PROCEDURE

Weigh out approximately 1.00 g (from 0.90 to 1.20 g) of aluminum foil or clean, thin scrap aluminum to the nearest 0.01 g. Cut it into small pieces, and put these in a 250-mL beaker. Add 50 mL of 1.5 M potassium hydroxide, KOH, solution.

CAUTION! DO NOT SPLATTER THE SOLUTION

Heat the beaker gently on a hot plate with a glass rod in the beaker. Do not use a flame. Hydrogen gas will be formed, so the experiment must be carried out under a hood. After 5 to 10 minutes, the aluminum should be dissolved, except for small amounts of impurities. Heat until the fizzing around each piece stops and the liquid level is reduced to **just over half** the original volume. If the liquid level goes below half, add distilled water as needed. Remove the beaker from the heat and use a glass stirring rod to remove any bits of remaining solids. Filter the hot liquid, through a *thin* layer of glass wool or loose cotton the size of a button.

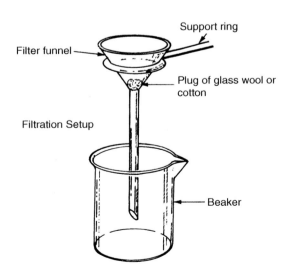

Support ring

Filter funnel

Plug of glass wool or cotton

Filtration Setup

Beaker

Cool the beaker of solution by running tap water over the *outside* of the beaker. The clear solution is then made acidic by the *slow* addition of 20.0 mL of 9 M H_2SO_4. The acidification should be accompanied with continuous stirring. After the sulfuric acid has been added, the solution may contain large lumps of aluminum hydroxide.

$$2.\ 2K^+ + 2Al(OH)_4^- + H_2SO_4 \longrightarrow 2K^+ + 2Al(OH)_3(s) + 2H_2O + SO_4^{2-}$$

This reaction mixture is then heated, gently, for about 5 minutes or until all of the aluminum hydroxide dissolves.

$$3.\ 2Al(OH)_3 + 3H_2SO_4 \longrightarrow 2Al^{3+} + 3SO_4^{2-} + 6H_2O$$

After the solution has become clear, remove it from the heat, filter (only if solid is present), and cool by setting the beaker first in tap water then in an ice bath while gently stirring for about 10 minutes until granular crystals form. Scratch the sides of the beaker with a glass rod to help induce crystallization. Set up the Buchner funnel assembly (see drawing). Place a circle of filter paper over the holes in the funnel, dampen it, and turn on the water to start the suction. Crystals are collected on the filter paper in the Buchner funnel using vacuum filtration. Gently dragging the policeman through damp crystals helps them dry. Wash crystals with 20 mL of a 50/50 alcohol-water mixture (in which alum is not very soluble). Pour the 50/50 mixture over the crystals on the filter paper, and dry them by additional suction. Weigh the crystals in a preweighed 100 mL beaker, and record the mass. The difference between these masses gives the "Grams of Product" for your label.

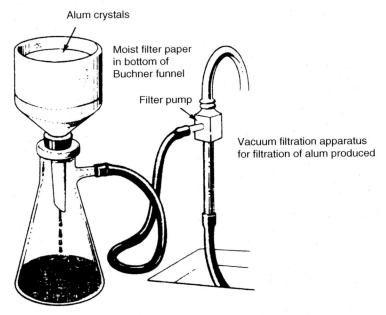

Alum crystals

Moist filter paper in bottom of Buchner funnel

Filter pump

Vacuum filtration apparatus for filtration of alum produced

Melting Point Determination:

Obtain a melting point capillary tube closed at one end and a small rubber band from the instructor. (If the tube is not already closed at one end, gently heat one end of the capillary tube in the edge of the flame of the Bunsen burner until the end completely closes. The end that is heated should be higher than the open end during the heating process. This prevents water from collecting in the closed end of the capillary. Rotate the tube between your thumb and forefinger during heating to prevent it from bending.)

Pulverize a small quantity of the alum sample with a clean mortar and pestle or use the round end of a test tube with crystals on a watch glass. Carefully push the open end of the capillary tube into the powdered sample, forcing a portion of the sample into the tube. Invert the tube and tap it on the lab bench to vibrate the sample into the closed end, or drop the capillary down a thermometer case onto the table. Repeat until the sample fills the tube to a depth of about 0.5 cm. Do not press too much sample into the capillary tube at one time or it will pack and then it cannot be shaken to the bottom of the tube.

Place the small rubber band 3 to 5 cm above the bulb of the lower end of a thermometer. Insert the capillary tube under the rubber band with the closed end and sample near the thermometer bulb. Insert the thermometer in a thermometer clamp or a split cork held in a utility clamp. Place the thermometer and capillary tube in a 400-mL beaker half to two-thirds full of tap water as pictured. Be sure that the open end of the capillary tube is above the water level.

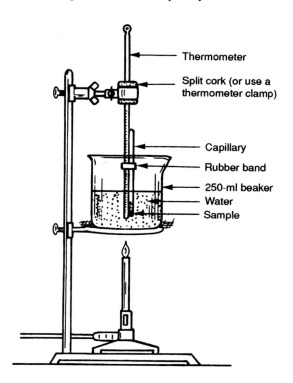

Heat the water slowly using a wire gauze, ring, and burner. Stir gently with a glass rod. A temperature increase of about 1 degree each 20 seconds is recommended. Observe the solid in the capillary very carefully. At the moment the solid starts to melt, read the temperature. This is the melting point of the solid. Record this temperature on your label and in your data.

Calculations: The reactions for the preparation of alum are summarized as follows:

4. Reaction of aluminum metal with potassium hydroxide:

$$2Al(s) + 2K^+ + 2OH^- + 6H_2O \longrightarrow 2K^+ + 2Al(OH)_4^- + 3H_2(g)$$
in solution

5. Acidification with sulfuric acid (summation of reactions 2 and 3):

$$2K^+ + 2Al(OH)_4^- + 4H_2SO_4 \longrightarrow 2K^+ + 2Al^{3+} + 4SO_4^{2-} + 8H_2O$$

6. Crystallization upon cooling:

$$2K^+ + 2Al^{3+} + 4SO_4^{2-} + 12H_2O \longrightarrow 2KAl(SO_4)_2 \bullet 12H_2O(s)$$
$$\text{Alum.}$$

Based on the balanced equations, one mole of aluminum produces one mole of alum. Calculate the theoretical yield of alum (in grams) produced from your scrap aluminum.

Theoretical yield of alum =

$$\underbrace{\frac{\text{g of Al}}{\text{sample}}}_{\text{grams Al} \longrightarrow \text{moles Al}} \times \underbrace{\frac{\text{1 mole Al}}{\text{27.0 g Al}}}_{} \times \underbrace{\frac{\text{1 mole alum}}{\text{1 mole Al}}}_{\substack{\text{moles} \\ \text{Al}} \longrightarrow \substack{\text{moles} \\ \text{alum}}} \times \underbrace{\frac{\text{molar mass } KAl(SO_4)_2 \bullet 12H_2O \text{ in g}}{\text{1 mole alum}}}_{\text{moles alum} \longrightarrow \text{grams alum}}$$

Then, calculate your percent yield:

$$\% \text{ yield} = \frac{\text{Mass of alum actually recovered}}{\text{Theoretical yield of alum in grams}}$$

Turn in your remaining sample in a 100 mL beaker or a container obtained from the instructor. Include the following items on the label.

Name of Compound: _____ Alum _____	Percentage Yield = _____ %	
Grams of product recovered = _____ g	Melting Point = _____ ° C	
Theoretical yield in grams = _____ g	Your Name _____	

SAMPLE CALCULATIONS:

Suppose you used 1.32 g of scrap aluminum and recovered 18.7 g of crystalline $KAl(SO_4)_2 \bullet 12H_2O$. Calculations are shown here.

$$\text{Molar mass of Al} = 27.0 \text{ g/mole}$$
$$\text{Molar mass of } KAl(SO_4)_2 \bullet 12H_2O = 474 \text{ g/mole}$$
Theoretical yield of alum, $KAl(SO_4)_2 \bullet 12H_2O$ from 1.32 g Al =

$$\underbrace{\frac{\text{1.32 g Al}}{\text{sample}}}_{\text{g Al} \longrightarrow \text{mole}} \times \underbrace{\frac{\text{1 mole Al}}{\text{27.0 g Al}}}_{} \times \underbrace{\frac{\text{1 mole alum}}{\text{1 mole Al}}}_{\substack{\text{moles} \\ \text{Al}} \longrightarrow \substack{\text{moles} \\ \text{alum}}} \times \underbrace{\frac{\text{474 g alum}}{\text{mole alum}}}_{\text{moles} \longrightarrow \text{g alum}} = \begin{array}{l} \text{23.2 g Alum} \\ \text{theoretical yield} \end{array}$$

$$\% \text{ yield} = \frac{18.7 \text{ g alum actually recovered}}{23.2 \text{ g alum theoretical yield}} \times 100\% = 80.6\% \text{ yield}$$

Report Sheet ______________________ ______________ ____________ ________

Name | Instructor/Section | Lab Partner | Date

Experiment 14: Alum

1. Mass of scrap aluminum used (to the nearest 0.0l g) ______________ g

2. Theoretical yield of alum in grams (see calculations below) ______________ g

3. Actual yield (grams of product recovered) to the nearest 0.0l g ______________ g

4. Percent yield (see calculations below) ______________ %

5. Melting point of alum ______________ °C

Show Calculations:

Theoretical yield calculations.

______________ g

Percent yield calculations.

______________ %

CONCLUSION

PROBLEMS TO DO

1. Suppose that you had started with 2.56 g of aluminum. What would be the theoretical yield of alum?

______________ g

2. Based on the theoretical yield calculations in the previous problem, what would be the percent yield if 44.3 g of alum is actually obtained?

______________ %

Experiment 14: PRE-LAB or POST-LAB QUIZ

Multiple choice questions. Circle your answers.

1. When the aluminum foil sample and potassium hydroxide are heated on the hot plate, continue heating until the aluminum dissolves and the solution

   a. boils down to one-fourth the original volume.
   b. boils down to one-half the original volume.
   c. boils down to three-fourths the original volume.
   d. boils completely dry.

2. Calculate the theoretical yield of alum that can be produced by using 1.08 g of aluminum, Al, using the information given here. (The molar mass of alum is 474 g/mol)

$$\text{mass of Al} \times \frac{1 \text{ mol Al}}{27.0 \text{ g Al}} \times \frac{1 \text{ mol alum}}{1 \text{ mol Al}} \times \frac{474 \text{ g alum}}{\text{mole alum}} = \text{mass of alum}$$

   Answer _____

3. Calculate the percentage yield of alum if 16.4 g of alum crystals formed when the theoretical yield was 19.7 g.

$$\% \text{ yield} = \frac{\text{mass of alum produced}}{\text{theoretical yield}} \times 100\% =$$

GRAVIMETRIC ANALYSIS OF A SOLUBLE SULFATE

PURPOSE

To determine the percent by mass of Na_2SO_4 in an unknown sample.

PERFORMANCE OBJECTIVE

Following this experiment, you should be familiar with the gravimetric approach for analysis of a sample.

BACKGROUND EXPLANATION

For nearly 200 years the technique of gravimetric analysis—a technique which involves weighing samples and their reaction products—has served as a highly accurate laboratory procedure for the determination of the percent composition (by mass) of a material. In this experiment, a gravimetric method will be used to determine the amount of Na_2SO_4 in a mixture where Na_2SO_4 is the only source of sulfate anion, SO_4^{-2}. The objective is to determine the percent of sodium sulfate (by mass) in the mixture to four significant figures.

The chemistry of this analysis is quite simple. The sample is weighed and then dissolved in distilled water and the sulfate anions are precipitated from solution as barium sulfate, $BaSO_4$, by the addition of a barium chloride, $BaCl_2$, solution.

$$BaCl_2 + Na_2SO_4 \longrightarrow BaSO_4(s) + 2\,NaCl$$

The net ionic equation is written as follows:

$$Ba^{+2}(aq) + SO_4^{-2}(aq) \longrightarrow BaSO_4(s)$$

The quality of the precipitate is improved if the precipitation is done in an acidified solution, so a small amount of hydrochloric acid, HCl, is added to the sample before the precipitation is begun. A better precipitate is also obtained if the solution is hot at the time of precipitation.

The precipitate is recovered by filtration, the filter paper is burned away from the precipitate (a technique called ignition), and the dried precipitate is weighed. During the analysis, sodium sulfate in the original sample is transformed to an equivalent amount of insoluble barium sulfate which is isolated away from the other constituents of the original sample. It is possible to calculate the amount of sodium sulfate that would have been needed to form the precipitated barium sulfate by using the balanced chemical equation:

$$BaCl_2 \ + \ Na_2SO_4 \ \longrightarrow \ BaSO_4(s) \ + \ 2\,NaCl$$

If we divide the mass of sodium sulfate calculated to be present in the original sample by the mass of the original sample, and multiply the result by 100%, we will obtain the percent (by mass) of sodium sulfate in the original sample.

PROCEDURE

Weigh out a sample of "soluble sulfate," the unknown, on the electronic analytical balance. To do so, precisely weigh a clean, dry 250 mL beaker and add 0.30 to 0.35 grams of the soluble sulfate and record the exact mass of the sample and the beaker to four decimal places. By subtracting, obtain the mass of the sample actually taken for analysis and record this mass.

Add about 50 mL of distilled water to the sample in the beaker. Add 10 drops of concentrated HCl solution (**CAUTION**) to the beaker of the dissolved sample. Stir the solution to dissolve the sample and leave the stirring rod in the beaker.

Using a clean graduated cylinder measure out 25 mL of 0.1M barium chloride, $BaCl_2$, solution. Heat the beaker with the sample solution until it is nearly—but not quite—boiling. Turn off the burner and *slowly* pour the barium chloride solution into the beaker of sample solution while stirring the hot liquid. Take about three minutes to add the barium chloride solution. Stop stirring, and wash the stirring rod with a small stream of distilled water into the beaker. Allow the beaker to sit undisturbed for 20 minutes while the precipitate settles.

While the precipitate settles, set up a funnel and ashless filter paper for filtration. Put a 400 mL beaker under the funnel to catch the filtrate (the liquid that passes through the filter). Pour some of the liquid from the beaker containing the barium sulfate into the filter. Pour it down the stirring rod and at no time fill the filter paper cone more than two-thirds full. Allow each portion to drain before adding more to the filter. When only a small portion of liquid remains in the beaker, swirl the beaker and transfer the precipitate over with the remaining liquid. Any precipitate remaining in the beaker should be washed from the walls into the filter using distilled water from a squeeze bottle. After the precipitate has been transferred, wash the precipitate with several small portions of distilled water. Let the filter drain completely.

While the filter drains, prepare a crucible. Heat the crucible in the hottest part of the Bunsen burner flame for about two minutes. Allow the crucible to cool on a ceramic pad on the bench top. When the crucible is cool, weigh the crucible on the same analytical balance used previously. Record the mass to four decimal places.

Now, with great care—so as not to tear the filter paper—remove the cone of the paper. Press the top edges together, fold them over, and carefully refold into a compact package that will fit into the crucible. **DO NOT** squeeze the precipitate itself to avoid rupturing the paper. Place the folded paper in the crucible.

Support the crucible with a clay triangle. Have the crucible positioned about 3-to-4 centimeters from the top of the burner.

For the first stage of heating, do not use the crucible cover. Heat the crucible with a *gentle* flame to dry the filter paper. Continue to heat moderately as the paper chars. If the paper bursts into flame, immediately smother the fire with the crucible cover and move the burner away to cool the crucible somewhat. Continue to heat until all the paper appears to have been converted to charcoal (i.e., carbon).

Heat the black char of carbon more strongly. The carbon will react with oxygen in the air to form gaseous carbon dioxide which escapes. Slowly, the residue will turn white. The white residue is the barium sulfate. The filter paper used in this analysis is the "ashless" variety. It leaves no weighable residue. Finally, heat the crucible enough to get the bottom red hot for several minutes and then allow the crucible to cool for ten minutes.

Using the same balance as before, weigh the crucible and its contents. Record this mass to four decimal places. Subtract out the mass of the crucible to obtain the mass of the barium sulfate. Discard the barium sulfate in the trash can (it is not hazardous).

Using the balanced chemical equation, calculate the mass of sodium sulfate in the original sample. (Notice that these two chemicals are in a one-to-one mole ratio.) Then, calculate the percent of sodium sulfate in the original sample. Record your values on the data sheet. Show your set up to calculate the mass of sodium sulfate. Be sure to complete the problems on the report sheet.

Experiment 15: Gravimetric Analysis

1. Mass of 250 mL beaker   _____ g

2. Mass of 250 mL beaker and sample   _____ g

3. Mass of soluble sulfate sample alone   _____ g

4. Mass of crucible alone   _____ g.

5. Mass of crucible and barium sulfate   _____ g

6. Mass of barium sulfate alone   _____ g

7. Molar mass of barium sulfate, $BaSO_4$ _____ g/mole

8. Moles of barium sulfate produced   _____ mole

9. Moles of sodium sulfate in sample   _____ mole

10. Molar mass of sodium sulfate, Na_2SO_4 _____ g/mole

11. Calculated mass of Na_2SO_4 in original sample   _____ g

12. Percent of Na_2SO_4 (by mass) in original sample (11÷3) _____ %

    Show calculations for the mass of sodium sulfate based on the mass of barium sulfate using the balanced equation.

PROBLEMS TO DO

1. What is the molarity of a solution of barium chloride if 67.3 grams of the salt barium chloride must be dissolved in enough water to give 2.50 liters of this solution? Hint: Molarity is in moles/liter.

2. How many grams of aluminum sulfate could be formed from 5.29 g of sodium sulfate with excess aluminum chloride?

$$2 \, AlCl_3 \; + \; 3 \, Na_2SO_4 \longrightarrow \; Al_2(SO_4)_3 \; + \; 6 \, NaCl$$

Experiment 15: PRE-LAB or POST-IAB QUIZ

Multiple choice questions. Circle your answers.

1. In this experiment the original sample contains sodium sulfate which reacts to form a
    precipitate having the chemical formula _____ .

2. If 0.311 g of an unknown sample for analysis is found to contain 0.225 g of
    Na_2SO_4, what is the percent of Na_2SO_4 in the sample?

    Show your work:

THE PRESSURE—VOLUME RELATIONSHIP FOR GASES

PURPOSE

To study the relation between the volume of a sample of gas and pressure.

PERFORMANCE OBJECTIVE

After doing this investigation, you should be familiar with the pressure-volume relationship for a gas and the graphical representation of this phenomenon.

MATERIALS

"J" shaped glass tube closed at the short end, meter stick, liquid mercury, and barometer or a video with appropriate data.

PROCEDURE

View a video of this investigation, or assemble a "J" shaped tube (closed at the short end) clamped in an upright position and add a small amount of mercury. Shake the tube to level the mercury. Record the volume of entrapped gas and the barometric pressure. Add mercury, then record volume and pressure for ten or more different readings.

DATA: see next page.

GRAPH

Prepare a full page graph of the total pressure versus total volume, with pressure on the ordinate (vertical) and volume on the abscissa (horizontal) axis. Label both axes (pressure and volume).

Circle points plotted and draw the best smooth curve that passes through or near these points. Describe the shape of the curve. Are P and V inversely or directly proportional?

CONCLUSION

Summarize your findings. What do you conclude when you observe the P × V column? Write a mathematical equation to represent this relation.

Barometric pressure	"J" tube upper reading (open end)	"J" tube lower reading (closed end)	Difference (upper minus lower)	* **Total pressure**	* **Volume of gas**	P × V

- The "J" tube upper reading is for the open end of the tube.
- The "J" tube lower reading is for the closed end of the tube.
- Total pressure = barometric pressure + difference between mercury levels.
- Plot points for total pressure and gas volume on the graph, making sure the graph FILLS the page.

Report Sheet ______________________ ____________ __________ ______

Name | Instructor/Section | Lab Partner | Date

Experiment 16: The Pressure–Volume Relationship

Experiment 16: POST-LAB QUIZ

Multiple choice questions. Circle your answers.

1. For a sample of gas at constant temperature, as pressure increases

    a. volume increases.
    b. volume decreases.
    c. volume remains constant.

2. For gases at constant pressure, gas pressure and volume

    a. are directly proportional.
    b. are inversely proportional.
    c. vary inconsistently and erratically.

DETERMINATION OF THE MOLAR VOLUME OF A GAS

PURPOSE

To experimentally determine the volume of a mole (L/mol) of hydrogen gas, and to use gas laws and correction factors as needed to convert the volume to standard conditions, STP.

PERFORMANCE OBJECTIVES

Following this investigation you should be able to (1) perform necessary calculations to determine the molar volume of a gas from experimental data, and to (2) perform calculations involving moles and correction factors for pressure and temperature.

MATERIALS

50 mL eudiometer, (gas measuring tube), battery jar or 2000 mL beaker, a two-hole stopper to fit eudiometer, 10 mL concentrated (12 M) HCl, about 5.5 cm Mg ribbon, and about one foot of extra fine copper wire.

BACKGROUND EXPLANATION

The reaction of magnesium metal with an excess of hydrochloric acid to produce magnesium chloride and hydrogen gas will be used to investigate the relation between moles of a gas and volume. The ratio, called the molar volume of the gas, has units in liters per mole.

$$Mg(s) + 2HCl(aq) \longrightarrow MgCl_2(aq) + H_2(g)$$

PROCEDURE   CAUTION: WEAR SAFETY GOGGLES

1. Obtain a 5.5 cm length of magnesium ribbon and polish it with steel wool (3 strokes) to remove any oxide coating. Then wipe off any loose particles. Weigh the Mg ribbon using an analytical balance to 0.0001 g.
2. Roll the Mg ribbon into a small ball and tightly wrap enough copper wire around it to cover it completely, but leave 2 to 3 inches of excess copper wire.

3. Using a funnel, carefully pour concentrated hydrochloric acid into the eudiometer (gas measuring tube) until it reaches the 10 mL mark. This provides an excess of HCl. With care, add tap water to the acid until the eudiometer is filled to overflowing. Care should be taken to prevent mixing.
4. Suspend the magnesium ball in the eudiometer by bending the end of the copper wire over the edge of the eudiometer and inserting the rubber stopper to hold the copper wire in place.
5. Hold your finger over the stoppered end of the eudiometer and invert it into a battery jar or 2000 mL beaker of water. Clamp the tube to hold it in a vertical position.
6. After the reaction is completed, adjust the water level inside and outside the tube so they are at equilibrium. To do this, raise or lower the eudiometer tube, or add water, but do not let the end of the eudiometer be raised above the water level. If this cannot be accomplished, ask your lab instructor for assistance. Carefully read and record the volume of gas produced, the water temperature, the gas temperature (put the thermometer into the eudiometer), and the barometric pressure.

Water Temp. (°C)	Vapor Pressure (mm Hg)	Water Temp. (°C)	Vapor Pressure (mm Hg)
16	13.6	23	21.1
17	14.5	24	22.4
18	15.5	25	23.8
19	16.5	26	25.2
20	17.5	27	26.7
21	18.7	28	28.3
22	19.3	29	30.0

Data

On the report sheet record the following:

a. The mass of magnesium.
b. The volume of hydrogen gas produced when water levels are equal inside and outside the tube.
c. The temperature of the water in the beaker.
d. The temperature of the hydrogen gas (put the thermometer in the eudiometer).
e. Barometric pressure (read the barometer in the lab).
f. Calculate moles of magnesium.
g. Calculate moles of hydrogen gas (same as moles of magnesium) based on the chemical equation for the reaction.
h. Calculate the volume of hydrogen gas collected, corrected to standard conditions, STP.
i. Calculate the molar volume of hydrogen gas at STP in L/mol.

Calculations

Using the mass of Mg, calculate moles of Mg, and moles of H_2 produced, based on the chemical equation. Using the temperature of H_2 gas, barometric pressure, vapor pressure of water, partial pressure of H_2, and volume of H_2 produced, calculate the vol. of H_2 corrected to STP, and liters H_2/mole at STP.

Name   Instructor/Section   Lab Partner   Date

Experiment 17: Molar Volume

Data

1. Mass of magnesium ribbon to nearest 0.0001 g   _____ g

2. The volume of hydrogen gas produced   _____ mL
   (Record this volume when water levels are equal
   inside and outside the eudiometer tube.)

3. Temperature of water in the beaker   _____ °C

4. Temperature of hydrogen gas   _____ °C

5. Barometric pressure in the lab.   _____ mm Hg

Calculations

We want to calculate the molar volume (liters per mole) for hydrogen gas at standard temperature and pressure, STP, so we must determine the number of moles of gas and the volume in liters, and then divide liters (Number 8) by moles (Number 7) to obtain the number of liters of H_2 per mole at STP (Number 9). This is called the molar volume.

Moles of hydrogen gas:

6. Moles of magnesium used

   Mass of Mg $\times \dfrac{1 \text{ mol Mg}}{24.3 \text{ g Mg}} =$ _____ mol Mg

7. Moles of hydrogen gas produced   _____ mol H_2
   (See the chemical equation; notice that moles
   of magnesium and hydrogen are in a 1 to 1 ratio.)

Volume of hydrogen gas produced (in liters), corrected to STP:

8. Calculate the milliliters of hydrogen gas that would be present at STP.
   Use the equation:

   Conditions at STP   Conditions in the laboratory

   $$\frac{P_{STP} \times V_{STP}}{T_{STP}} = \frac{P_{lab} \times V_{lab}}{T_{lab}}$$

   $$\frac{760 \text{ torr} \times V_{STP}}{273 \text{ K}} = \frac{(P_{barometer} - P_{water\ vapor}) \times V_{gas\ collected}}{\text{Gas temperature in K}}$$

   Note: To find the vapor pressure of water ($P_{water\ vapor}$) locate the water temperature in the table of vapor pressures and read the appropriate vapor pressure in the adjacent column.

   Rearrange the equation and solve for V_{STP}. $V_{STP} = \dfrac{(\quad)(\quad)(\quad)}{(\quad)(\quad)} = $ _____ mL = _____ L

9. The MOLAR VOLUME of hydrogen at STP   _____ L/mol
   (Number 8 divided by Number 7)

Name   **Instructor/Section**

Experiment 17: PRE-LAB or POST-LAB QUIZ

Multiple choice questions. Circle your answers.

1. Based on the chemical reaction of Mg with HCl(aq), each mole of Mg reacts to give

    a. one-half mole of hydrogen gas.
    b. one mole of hydrogen gas.
    c. two moles of hydrogen gas.
    d. three moles of hydrogen gas.
    e. one mole of chlorine gas.

2. Before reading the volume of gas produced by the reaction, the eudiometer (the gas measuring tube) must be raised or lowered so the water level inside the tube is

    a. HIGHER than the water level outside the tube.
    b. LOWER than the water level outside the tube.
    c. IN LINE WITH the water level outside the tube.
    d. EXACTLY AT the 50.0 mL mark.

3. The final mathematical step in determining the molar volume of a gas, is to

    a. divide liters of gas by moles of gas produced.
    b. multiply liters of gas by moles of gas produced.
    c. divide moles of gas by liters of gas produced.
    d. multiply moles of gas by liters of gas produced.
    e. divide grams of Mg used by milliliters of gas produced.

4. The gas pressure inside the collecting tube is from

    a. hydrogen gas only.
    b. hydrogen gas and oxygen gas.
    c. hydrogen gas and chlorine gas.
    d. hydrogen gas and water vapor.
    e. hydrogen gas, chlorine gas, and water vapor.

DIFFUSION, THE RACE BETWEEN GASES

18

PURPOSE

To investigate diffusion of gases; that is, movements of gases at a constant temperature and pressure.

PERFORMANCE OBJECTIVES

After completing these investigations, you should be able to relate the process of diffusion to the kinetic theory of gases. This study will allow you to investigate the relation of gas movement to molar mass, so you can apply Graham's Law using simple laboratory apparatus. Subsequently, you should be able to apply Graham's Law, when needed, in solving problems.

BACKGROUND EXPLANATION

The kinetic theory of gases provides an explanation for the movement of gas molecules. These molecules move with high velocities or speed, but their motion is random and chaotic, so the rate of gas movement through space (diffusion) is decreased by frequent collisions with other gas molecules. The rate of diffusion is proportional to, but not equal to, the velocity of the molecule.

According to Avogadro's Hypothesis, equal volumes of gas at the same temperature and pressure contain equal numbers of molecules. These molecules, then, must exert an equivalent amount of force on the container walls. The kinetic theory states that all kinds of gas molecules have the same average kinetic energy, at the same conditions. Since kinetic energy is dependent upon mass and velocity (K.E. = $\frac{1}{2} mv^2$), it follows that the velocity of the molecule is dependent on the mass of the molecule.

The experimental investigation outlined here is designed to demonstrate the relation between diffusion and molar mass. When two gases are at the same pressure and temperature, their kinetic energies are equivalent, so

$$\frac{1}{2} m_A u_A^2 = \frac{1}{2} m_B u_B^2$$

where A and B represent different gases, each with mass, m, and velocity, u.

Multiplying by 2 and rearranging terms, gives the relation:

$$\frac{u_A^2}{u_B^2} = \frac{m_B}{m_A}$$

Because the mass is proportional to the molar mass, the m_B/m_A ratio may be replaced by the ratio of molar masses M_B/M_A.

The relation then becomes:

$$\frac{u_A{}^2}{u_B{}^2} = \frac{M_B}{M_A} \quad \text{or, by squaring} \quad \frac{u_A}{u_B} = \frac{\sqrt{M_B}}{\sqrt{M_A}}$$

We can broaden the relation to include distance and time. The rate, u, is proportional to distance, s, but inversely proportional to time, t and the square root of molar mass. The expanded relation is

$$\frac{u_A}{u_B} = \frac{s_A}{s_B} = \frac{t_B}{t_A} = \frac{\sqrt{M_B}}{\sqrt{M_A}}$$

The two gases selected for this experiment, ammonia and hydrogen chloride, HCl, react to form the white compound NH_4Cl at room temperature. It will be easy to detect the white ring formed when these gases travel through a tube and meet. Since we can readily measure this distance experimentally, we shall investigate the relation of distance traveled, s, and molar mass, M. From the preceding equations we select the relation:

$$\frac{s_A}{s_B} = \frac{\sqrt{M_B}}{\sqrt{M_A}}$$

MATERIALS

Glass tube, about 70-80 cm with an inside diameter of 12-15 mm, 2 cotton balls, two 100–mL beakers, meter stick, corks, buret clamp support, concentrated ammonia solution, and concentrated hydrochloric acid.

CAUTION: WEAR GOGGLES; HANDLE THESE CHEMICALS WITH CARE.

PROCEDURE

This experiment should be done under a hood, if possible, to reduce lab pollution. Secure the 70 to 80 cm dry glass tube in a horizontal position using a ring stand and a buret clamp. Pour out small amounts of concentrated HCl and concentrated ammonia into separate beakers. Keep these apart to minimize the formation of NH_4Cl. The concentrated ammonia is used as a source of NH_3 gas.

Use forceps to dip the two cotton balls into the two liquids, separately. Saturate the surface of each ball with the appropriate liquid, place them, simultaneously, at opposite ends of the glass tube, and secure the cotton balls in place. Record the initial and ending times.

It may take 5 to 10 minutes before the white NH_4Cl ring forms in the tube where the two gases meet. Look carefully to find the foggy white ring to the right or left of center. Measure and record the distance, in centimeters, that each gas travels.

Determine the ratio of distance traveled for HCl to ammonia and compare this to the predicted ratio.

Experiment 18: Diffusion

Diffusion Data

1. Distance HCl gas diffuses to form the "white ring"   _____ cm

2. Distance ammonia gas diffuses to form the "white ring"   _____ cm

3. Time at beginning of experiment   _____

4. Time at which white ring is first observed   _____

5. Elapsed time   _____ min

Calculations

6. Molar mass (MM) of HCl based on the periodic table   _____ g/mol

7. Molar mass (MM) of ammonia (NH_3) based on the
   periodic table   _____ g/mol

8. Ratio of HCl distance traveled to ammonia distance traveled
   (line 1 divided by line 2)   _____

9. Predicted ratio of HCl to ammonia distance based on
   molar masses

   ratio of HCl to ammonia distance $= \dfrac{\sqrt{MM}_{ammonia}}{\sqrt{MM}_{HCl}}$ _____

10. Percent error:

   $\dfrac{\text{Difference between experimental and predicted ratio}}{\text{predicted ratio}} \times 100\% =$ _____

11. Rate of diffusion of HCl $\dfrac{\text{Line 1}}{\text{Line 5}} = \dfrac{cm}{min}$ _____ cm/min

12. Rate of diffusion of ammonia $\dfrac{\text{Line 2}}{\text{Line 5}} = \dfrac{cm}{min}$ _____ cm/min

Conclusion: Compare the relative rates of diffusion for a gas having a low molar mass with a gas having a high molar mass.

Experiment 18: PRE-LAB or POST-LAB QUIZ

Multiple choice questions. Circle your answers.

1. The rate at which a gas travels is

    a. DIRECTLY proportional to the distance traveled.
    b. INVERSELY proportional to the distance traveled.
    c. NOT RELATED to the distance traveled.

2. The distance a gas travels is

    a. DIRECTLY proportional to the square root of its molar mass.
    b. DIRECTLY proportional to its molar mass squared.
    c. INVERSELY proportional to the square root of its molar mass.
    d. INVERSELY proportional to its molar mass squared.
    e. NOT RELATED to its molar mass.

3. The rate at which a gas travels (or diffuses) is

    a. DIRECTLY proportional to the square root of its molar mass.
    b. DIRECTLY proportional to its molar mass squared.
    c. INVERSELY proportional to the square root of its molar mass.
    d. INVERSELY proportional to its molar mass squared.
    e. NOT RELATED to its molar mass.

4. Which one of the following gases has the greatest rate of diffusion?

    a. O_2 b. N_2 c. F_2 d. Cl_2

HEAT OF NEUTRALIZATION

19

PURPOSE

To determine enthalpies or heats of reaction using styrofoam cups as calorimeters, and to study Hess's law.

PERFORMANCE OBJECTIVES

Following these investigations, you should be familiar with calorimetry. You should be able to calculate enthalpies of solution and neutralization and to apply Hess's law.

MATERIALS

Two styrofoam cups stacked together as a calorimeter and a piece of cardboard or one styrofoam cup trimmed one-half inch from the top and placed upside down in the others for a lid, 0.1° C thermometer, stirrer, chemicals, and plastic weigh boat for NaOH pellets.

BACKGROUND EXPLANATION

In this experiment you will use a styrofoam cup both as the reaction vessel and as a simple calorimeter to measure the heat evolved or absorbed during the reactions.

You may assume that the enthalpy or heat of reaction will be used to change the temperature of the aqueous solution only. Neglect small losses of heat to the surroundings. Recall that it takes one calorie of energy to change the temperature of one gram of water one degree Celsius. This is the specific heat of water, 1.00 cal/gram °C.

You do not need to weigh the water used because the mass of one milliliter of water is one gram. When the reactants are added to your calorimeter you should record the change in temperature to the nearest 0.1 °C. From the change in temperature, ΔT, and the mass of the reactants you can calculate the number of calories evolved or absorbed, Q.

$$\text{mass} \times \Delta T \,^{\circ}C \times \text{specific heat} = Q$$

In this experiment you will measure and compare the quantity of the heat involved in three reactions:

Reaction 1: Solid sodium hydroxide dissolves in water to form an aqueous solution of ions:

$$NaOH(s) \longrightarrow Na^+(aq) + OH^-(aq) + X_1 \text{ cal}$$
$$\Delta H_1 = -X_1 \text{ cal}$$

Reaction 2: Solid sodium hydroxide reacts with an aqueous solution of hydrogen ions and chloride ions to form water and an aqueous solution of sodium chloride:

$$NaOH(s) + H^+(aq) + Cl^-(aq) \longrightarrow H_2O + Na^+(aq) + Cl^-(aq) + X_2 \text{ cal}$$
$$\Delta H_2 = -X_2 \text{ cal}$$

Reaction 3: An aqueous solution of sodium hydroxide reacts with an aqueous solution of hydrogen ions and chloride ions to form water and an aqueous solution of sodium chloride:

$$Na^+(aq) + OH^-(aq) + H^+(aq) + Cl^-(aq) \longrightarrow H_2O + Na^+(aq) + Cl^-(aq) + X_3 \text{ cal}$$
$$\Delta H_3 = -X_3 \text{ cal}$$

Procedure

1. To determine the heat of Reaction 1:

   a. Pour 150. mL (±1 mL) of cool tap water into the cup. Stir carefully with a thermometer until a constant temperature is reached (about room temperature). Record this temperature to the nearest 0.1° C.

   b. Weigh about 1.6 g of solid sodium hydroxide, NaOH, to the nearest 0.01 g using a weighing boat. Sodium hydroxide becomes moist in the open air, so make weighings quickly. Do not touch the NaOH; it is caustic.

   c. Pour the weighed NaOH(s) into the water in the cup. Place the thermometer into the cup, cover it, stir, and record the highest temperature reached. (To avoid erratic readings, the thermometer should not touch solid NaOH pellets.)

   d. Before proceeding to Reaction 2, rinse the cup with tap water.

2. To determine the heat of Reaction 2:

   e. Repeat steps a, b and c above except in step (a) substitute 160. mL of 0.25 M HCl for tap water.

   f. Rinse the cup again and proceed to Reaction 3.

3. To determine the heat of Reaction 3:

   g. Measure out into a beaker 80. mL of 0.5 M NaOH. Place 80. mL of 0.5 M HCl in the calorimeter. Be sure that both of these solutions are slightly below room temperature. You should check the temperature of each solution with a thermometer being careful to rinse and dry it before transferring it from one solution to the other. Record the temperatures of both solutions. The average of these temperatures may be taken as the initial temperature. Pour the sodium hydroxide solution into the hydrochloric acid solution. Mix quickly and record the highest temperature reached.

Experiment 19: Heat of Neutralization

	Reaction 1	Reaction 2	Reaction 3
1. Initial temperature			
2. Final temperature			
3. The change in temperature, ΔT			
4. Heat, Q, absorbed by the solution = mass of soln. $\times \Delta T \times 1$ cal/g °C Note: Mass = sum of masses of solids and liquids used.			
5. The amount of heat absorbed by the cup (neglect this quantity)	neglect	neglect	neglect
6. TOTAL amount of heat, Q, absorbed			
7. The number of moles of NaOH used in each reaction For solid: gNaOH $\times$ 1/MM NaOH For solution: L NaOH $\times$ Molarity			
8. The heat of reaction per mole of NaOH, ΔH. Include sign. (Line 6/Line 7)			

Questions

1. Write the net ionic equations for reactions 1, 2, and 3.

2. In reaction 1, ΔH_1 represents the heat of solution of NaOH(s). Look at the net ionic equations for reactions 2 and 3 and make a statement concerning the significance of ΔH_2, that is, what has happened to produce the energy change. Also explain the meaning of ΔH_3.

3. How does ΔH_2 compare with the sum of $\Delta H_1 + \Delta H_3$. Explain.

4. Calculate the percent difference between ΔH_2 and the SUM of $\Delta H_1 + \Delta H_3$, using ΔH_2 as the reference.

$$\frac{\text{Difference}}{\Delta H_2} \times 100\% =$$

_____ % difference

137

5.   Suppose you had used 4 g of NaOH(s) in reaction 1 instead of 1.6 g.
    a) What would be the number of calories evolved?

    b) What effect would using 4g NaOH have on your calculation of ΔH_1, the heat evolved
       per mole of NaOH in line 8?

Conclusion

Experiment 19: PRE-LAB or POST-LAB QUIZ

Given the following three chemical equations labeled A, B, and C (not necessarily in the order listed in the lab manual):

A. $NaOH(s) \rightarrow Na^+(aq) + OH^-(aq)$

B. $Na^+(aq) + OH^-(aq) + H^+(aq) + Cl^-(aq) \rightarrow H_2O + Na^+(aq) + Cl^-(aq)$

C. $NaOH(s) + H^+(aq) + Cl^-(aq) \rightarrow H_2O + Na^+(aq) + Cl^-aq)$

Multiple choice questions. Circle your answers.

1. Heat released (or absorbed) during the neutralization of sodium hydroxide solution by hydrochloric acid is represented by

   a. equation A.
   b. equation B.
   c. equation C.

2. Heat released (or absorbed) when solid sodium hydroxide is dissolved is represented by

   a. equation A.
   b. equation B.
   c. equation C.

3. Heat released (or absorbed) when sodium hydroxide is both dissolved and neutralized is represented by

   a. equation A.
   b. equation B.
   c. equation C.

HEAT OF SOLUTION

20

PURPOSE

To determine the heat of solution in cal/g for a soluble unknown compound.

PERFORMANCE OBJECTIVES

Following this investigation, you should be able to experimentally determine the enthalpy of solution for a soluble compoud.

MATERIALS

Two styrofoam cups stacked together as a calorimeter and a piece of cardboard or one styrofoam cup trimmed one–half inch from the top and placed upside down in the others for a lid, 0.1°C thermometer, stirrer, chemicals, and a plastic weigh boat for NaOH pellets.

PROCEDURE

The enthalpy of solution—or heat of solution—for a soluble solid chemical can be obtained by the following method described here. Weigh an empty styrofoam cup calorimeter to the nearest 0.01 g. Place about 50 mL of distilled water in the calorimeter and weigh it to the nearest 0.01 g. Subtract these masses to obtain the mass of water. Measure the initial temperature of water to the nearest 0.01°C. The temperature of the water should be slightly below room temperature.

Weigh out a 5 g sample of unknown to the nearest 0.001 g. Pour the compound into the calorimeter and record the maximum or minimum temperature to the nearest 0.1°C as the solid compound dissolves. A temperature change of 5°C or more should be observed. It may be necessary to repeat the experiment and adjust the amount of water or solid accordingly. Calculate the heat absorbed by the calorimeter, Q, using the equation shown here.

$$Q = Q_w + Q_s$$
$$Q = (m_w)(\Delta T_w)(1 \text{ cal/g } °C) + (m_s)(\Delta T_s)(0.2 \text{ cal/g } °C)$$

The heat absorbed by the calorimeter, Q, is the sum of the heat absorbed by the water Q_w plus the heat absorbed by the solid Q_s. For our purposes, the specific heat value of 0.2 cal/g°C for the solid chemical used is close enough. The sum of heat gained by the calorimeter and heat lost by the solid is zero. The change in enthalpy in calories/gram equals $Q_{solution}$ divided by the mass of the chemical used. The change in enthalpy, ΔH, is negative for an exothermic process and positive for an endothermic process.

Experiment 20: Heat of Solution

Heat of solution for compound No. _____

1. Mass of empty calorimeter   _____ g

2. Mass of calorimeter with water   _____ g

3. Mass of water m_w _____ g

4. Mass of solid chemical m_s _____ g

5. Initial temperature T_1 _____ °C

6. Final temperature T_2 _____ °C

7. Temperature change $(T_2 - T_1)$ _____ °C

8. Q_w (see equation)   _____ cal

9. Q_s (see equation)   _____ cal

10. Total $Q = Q_w + Q_s$ *   _____ cal

11. Calculate the enthalpy change or heat
    of solution per gram of solid ** $\Delta H_{(Soln)}$ = _____ cal/g

* The Q calculated in line 10 represents the total Q, the heat absorbed or released by the calorimeter contents. We assume no heat loss due to the transfer of heat so heat of solution, ΔH_{Soln}, equals $-Q$.

$$** \ \Delta H_{(Soln)} = \frac{-Q_{total}}{\text{mass of chemical used}}$$

Notice that ΔH is negative for an exothermic process, and positive for an endothermic process, the opposite sign of Q.

Show calculations here.

Name ______________________ Instructor/Section ______________________

Experiment 20: PRE-LAB or POST-LAB QUIZ

Multiple choice questions. Circle your answers.

1. The amount of heat absorbed by the calorimeter contents is

   a. the DIFFERENCE between the heat absorbed by the water and heat lost by the chemical being dissolved.
   b. the SUM of the heat absorbed or lost by the water and heat absorbed or lost by the chemical being dissolved.
   c. assumed to be EQUAL to the heat absorbed by the water only.
   d. assumed to be EQUAL to the heat lost by the water only.

2. The amount of heat absorbed by the calorimeter and contents is

   a. equal in value—but opposite in sign—to the heat lost by the chemical being dissolved.
   b. equal—in value and in sign—to the heat lost by the chemical being dissolved.
   c. always more than is lost by the chemical being dissolved.
   d. sometimes more and sometimes less than the heat lost—only by experiment do we know.

3. If the heat of solution for a chemical was –22,000 cal/g when 2.0 g of solid was dissolved, the value for the heat of solution using 4.0 g should be

   a. –44,000 cal/g
   b. –22,000 cal/g
   c. –11,000 cal/g

PERCENTAGE OF ACETIC ACID IN VINEGAR

PURPOSE

To analyze commercial vinegar and determine the percentage of acetic acid using volumetric analysis.

PERFORMANCE OBJECTIVES

Following this vinegar analysis, you should be able to perform volumetric analysis on unknown acids using phenolphthalein as an indicator. You should be able to calculate the percentage by weight of the unknown from data obtained.

BACKGROUND EXPLANATION

Volumetric analysis is a very important and standard method of determining the concentrations of solutions. A measured volume of a solution of known concentration is reacted with just enough of the solution with an unknown concentration to cause a complete reaction. Completion of the reaction can be detected by some observable change. The volumetric analysis process is called **titration**.

To minimize errors, volumes of reacting liquids can be measured with great precision to the nearest 0.1 mL and estimated to the nearest 0.01 mL when burets are used. Become familiar with standard methods of cleaning, filling, reading, and operating a buret.

Some type of indicator is used to signal the point of complete reaction. When the reaction involves acid and base neutralization, a pH change can be observed with a pH meter or an indicator solution such as phenolphthalein, methyl orange, methyl red, and so on. These indicators are organic dyes that change color at specific pH intervals. The point at which the indicator changes color is known as the **end point**, the point at which the titration is stopped. If the correct indicator is chosen, then the end point will coincide with the **equivalence point**, which is the exact point when equivalent quantities have reacted. The chemical equation is

$$CH_3COOH + NaOH \longrightarrow NaCH_3COO + H_2O$$
$$\text{acetic acid} \qquad\qquad\qquad \text{sodium acetate}$$

We can see that one mole of acetic acid exactly neutralizes one mole of sodium hydroxide. From the data collected, we can calculate the percentage of acetic acid in a vinegar sample.

Calculations for the percentage of acetic acid in vinegar samples are outlined under the calculations section of the report sheet.

MATERIALS

50 mL buret, 10 mL pipet (or a second 50 mL buret), 250 mL Erlenmeyer flash, wash bottle with distilled water, 0.500 M standard NaOH, and vinegar.

PROCEDURE

Use a clean buret that has been washed with detergent, rinsed with tap water, and distilled water (two samples of 2 mL each). Rinse the clean buret with NaOH (1 to 2 mL). Set up a buret on a buret stand with a buret clamp. Fill the buret to above the 0.00 mL mark with the NaOH solution and let the solution drain out in an extra beaker until the meniscus is on or below the 0.00 mL mark. In a clean Erlenmeyer flask, carefully measure out about 10.0 mL of vinegar from a buret set up on the table where chemicals are dispensed (or use a 10-mL pipet to measure out the vinegar). Record the precise volume of vinegar obtained. Add about 15 mL of distilled water. The amount of water added is not critical because, at the equivalence point, the solution will be neutral, regardless of the amount of distilled water added. Also add 2 drops of phenolphthalein indicator to the flask containing the vinegar and water. Swirl to mix.

Titrate the vinegar with 0.500 M standard NaOH by adding the NaOH slowly, dropwise, to the vinegar solution with swirling action as shown.

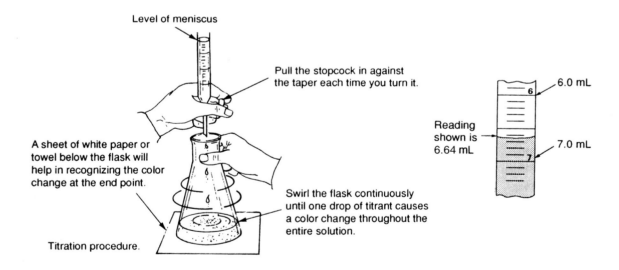

Level of meniscus

Pull the stopcock in against the taper each time you turn it.

A sheet of white paper or towel below the flask will help in recognizing the color change at the end point.

Swirl the flask continuously until one drop of titrant causes a color change throughout the entire solution.

Titration procedure.

Reading shown is 6.64 mL

6.0 mL
7.0 mL

As the end point is approached, the pink phenolphthalein color will be more persistent. Near the end point, add titrant (NaOH) *dropwise* with swirling. The end point is reached with the drop that causes the solution to turn pink and remain pink. Record the volume to the nearest 0.01 mL. Repeat the analysis with lab partners taking turns doing the titration. Record your data and calculate the percentage of acetic acid.

Report Sheet ______________________ ____________ ________ ______

Name | Instructor/Section | Lab Partner | Date

Experiment 21: Vinegar Analysis

Data

	Trial 1	Trial 2
1. Vinegar, FINAL reading of buret	________ mL	________ mL
2. Vinegar, INITIAL reading of buret	________ mL	________ mL
3. VOLUME of VINEGAR used (1 minus 2)	________ mL	________ mL
4. NaOH, FINAL reading of buret	________ mL	________ mL
5. NaOH, INITIAL reading of buret	________ mL	________ mL
6. VOLUME of NaOH used (4 minus 5)	________ mL	________ mL
7. Molarity of NaOH used (see label on bottle)	________ M	________ M

Calculation of Moles and Percent Acetic Acid

8. Moles of NaOH used ________ mol NaOH ________ mol NaOH
(Liters NaOH used × Molarity in mol/L)

9. Moles of acetic acid in sample ________ mol acid ________ mol acid
(The balanced chemical equation shows that 1 mol of NaOH neutralizes 1 mol of acetic acid, so moles of NaOH and acid, lines 8 and 9, are equal.)

10. Mass of acetic acid in the sample ________ g CH_3COOH ________ g CH_3COOH
(Moles acetic acid × molar mass of acetic acid, CH_3COOH, 60.0 g/mol as shown below)

Method: _____ mol acetic acid $\times \dfrac{60.0 \text{ g acetic acid}}{1 \text{ mol acetic acid}} =$ ______ g acetic acid, CH_3COOH

11. Mass of vinegar sample used ________ g vinegar ________ g vinegar
(You could weigh the flask before and after you put in the vinegar sample or, more simply, you can calculate the mass of vinegar used by multiplying its volume times its density. The density of the vinegar is 1.005 g/mL.) Sample calculations follow:

Method: _______ mL vinegar $\times \dfrac{1.005 \text{ g vinegar}}{1 \text{ mL vinegar}} =$ ______ g vinegar

12. **Percent by mass** of acetic acid, CH_3COOH, in the vinegar sample

Method: $\dfrac{\text{Mass of acetic acid, Line 10}}{\text{Mass of vinegar, Line 11}} \times 100\% =$ ________ % CH_3COOH

________ % CH_3COOH ________ % CH_3COOH

13. **Average percent** acetic acid by mass ________ % CH_3COOH

Experiment 21: PRE-LAB or POST-LAB QUIZ

Multiple choice questions. Circle your answers.

1. The process of carrying out an analysis where a measured volume of a chemical having an UNKNOWN concentration is reacted with a measured volume of a chemical having a KNOWN concentration is called

   a. the neutralization point.
   b. the equivalence point.
   c. the end point.
   d. a titration.

2. In volumetric analysis, the point at which the indicator changes color is called

   a. the neutralization point.
   b. the equivalence point.
   c. the end point.
   d. a titration.

3. In this experiment, one mole of base neutralizes

   a. one-half mole of acid.
   b. one mole of acid.
   c. two moles of acid.

4. Base of known concentration is added to the vinegar until the solution in the flask turns from

   a. pink to colorless.
   b. colorless to the first tinge of pink, which will quickly go away.
   c. colorless to the first tinge of pink that does not go away when swirled.
   d. colorless to a bright shade of pink

5. What is the name and chemical formula of the acid present in vinegar?

   Name: _____   Formula: _____

ACIDS, BASES, AND pH

22

PURPOSE

• To use a pH meter, a naturally occurring dye, and common indicators to determine the pH values of several substances.

• To observe changes in pH as an acid or base is added to solutions that are buffered and to solutions that are not buffered.

PERFORMANCE OBJECTIVES

Following these investigations, you should be able to use a pH meter or indicators to obtain the pH of a given solution. You should also be able to compare the effects of adding an acid or base to a solution that is buffered and to one that is not buffered.

MATERIALS

Part A: red cabbage, 400-mL beaker, hot plate, and distilled water.

Part B: solutions with pH values of 0 to 14 in test tubes.

Part C: pH meter, test tubes, medicine droppers, and samples to test, such as tap water, vinegar, household ammonia, fruit juices, shampoos and conditioners, diluted laundry detergents, and so on.

Part D: buffer solution, 0.1 M HCl, 0.1 M NaOH, and pH test paper strips.

BACKGROUND EXPLANATION

Some natural substances contain pigments that change color when the pH changes. One example of such a substance is a pigment found in red cabbage leaves. We will extract (remove) this pigment from the leaves of red cabbage and use this natural pH indicator to determine pH values of various solutions.

In Part B you will set up a series of solutions with known pH values that range from 0 to 14. When the cabbage pigment is added to the series of solutions with different pH values, a range of colors can be observed. You can use this reference set of solutions with a range of pH values and colors to determine the pH values of several solutions.

The pH of a solution is a measure of the hydrogen ion concentration in the solution. A pH value between 0 and 7 is acidic, a pH of 7 is neutral, and a pH value between 7 and 14 is basic. The closer the pH is to 0, the more acidic it is, and the closer to 14 it is, the more basic it is.

$$0 \quad 1 \quad 2 \quad 3 \quad 4 \quad 5 \quad 6 \quad 7 \quad 8 \quad 9 \quad 10 \quad 11 \quad 12 \quad 13 \quad 14$$

very acidic   neutral   very basic

The pH of blood in the body is held within its narrow limits of 7.35 to 7.45 by **buffers** in the blood. Buffers can maintain a relatively constant pH in a solution even when small amounts of acid or base are added. A buffer is made up of either a weak acid and its salt or a weak base and its salt.

PROCEDURE

Carry out these investigations in groups of 4 to 8 students.

Part A: PREPARATION OF A NATURAL INDICATOR

Tear a few leaves of red cabbage into small pieces and place them in a 400 mL beaker. Add 100 mL of distilled water to cover the leaves. Place the beaker of leaves and water on a hot plate and boil gently for ten minutes. The solution should have a deep purple color. If it doesn't, add more pieces of the cabbage leaves and boil again. Remove the beaker from the hotplate and allow it to cool or run cool tap water over the outside of the beaker for more rapid cooling.

While preparing the cabbage indicator solution, set up a series of 15 test tubes, label them in sequence with pH values of 0 to 14, and to each test tube add 4 mL of the solution having the appropriate pH. Try to keep the volumes in the test tubes about the same. This is your set of pH reference standards.

Part B: PREPARATION OF pH REFERENCE STANDARDS

Use a medicine dropper to consistently add 40 drops of the cabbage indicator to each of the solutions in the pH reference set and stir to agitate. Record the colors of the pH standards on your data sheet. Be descriptive! Keep the pH reference set available for comparison as needed in Part C.

Part C: DETERMINATION OF pH VALUES

Using Red Cabbage Indicator: Into a test tube measure out 4 mL of one of the solutions to be tested and add 40 drops of the red cabbage indicator. Stir to agitate the sample and compare the resulting color to the colors in your pH reference set. The pH of the reference solution having the closest match to the color of the sample tested is the pH of the test sample. Record the name of the sample being tested and the pH on your data sheet. Repeat this procedure for other unknowns to be tested. Also test any unknowns you may have brought with you for testing.

Using a pH Meter: After completing your pH determinations, your instructor will measure the pH of the unknowns using a pH meter. Record these values on your data sheet beside the pH values obtained using red cabbage indicator.

Part D: EFFECT OF BUFFERS ON pH

Pour 5 mL of distilled water into each of two test tubes. Into two *different* test tubes, pour 5 mL of buffer. Use pH indicator paper strips—*not* red cabbage indicator—to determine the pH of these solutions.

Add 5 drops of 0.1 M HCl to one of the test tubes containing water and add another 5 drops of the acid to one of the test tubes containing a buffer. Record the pH values on your data sheet. Using the remaining two test tubes of water and buffer, repeat the procedure using 5 drops of 0.1 M NaOH and record your results. Determine the change in pH for the buffered and nonbuffered solutions.

Experiment 22: Acids, Bases, and pH

Part B data: Preparation of pH Reference Standards

pH	Color	pH	Color
0	_____	8	_____
1	_____	9	_____
2	_____	10	_____
3	_____	11	_____
4	_____	12	_____
5	_____	13	_____
6	_____	14	_____
7	_____		

Part C data: Determination of pH Values

	Solution	Color	pH With Cabbage	pH With Meter	Solution is Acidic, Basic, or Neutral
1.	_____	_____	_____	_____	_____
2.	_____	_____	_____	_____	_____
3.	_____	_____	_____	_____	_____
4.	_____	_____	_____	_____	_____
5.	_____	_____	_____	_____	_____
6.	_____	_____	_____	_____	_____
7.	_____	_____	_____	_____	_____
8.	_____	_____	_____	_____	_____
9.	_____	_____	_____	_____	_____
10.	_____	_____	_____	_____	_____

Part D data: Effect of Buffers on pH

	Water	Buffer
1. pH, initial	_____	_____
2. pH after adding 5 drops 0.1 M HCl	_____	_____
3. pH change	_____	_____
4. pH, initial	_____	_____
5. pH after adding 5 drops 0.1 M NaOH	_____	_____
6. pH change	_____	_____

7.   Which solution(s) showed the greatest pH change? Why?

8.   Which solution(s) showed little or no pH change? Why?

ADDITIONAL QUESTIONS:

1.   What is the function of a buffer?

2.   Approximately where on the pH scale would you expect to find the pH of a sample of "acid rain"?

3.   Normally, the pH of blood in the human body is maintained in a very narrow range between 7.35 and 7.45. A patient having an acidic blood pH of 7.3 may be treated with a mild base such as $NaHCO_3$. Why should this treatment raise the pH of blood?

 ______________________ ______________________
Name Instructor/Section

Experiment 22: PRE-LAB or POST-LAB QUIZ

Multiple choice questions. Circle your answers.

1. A mixture of a weak acid and its salt or a weak base and its salt is required for a(n)

   a. acid-base indicator.
   b. buffer.
   c. titration.
   d. reference standard.

2. A solution with a pH of 6.6 is

   a. very acidic
   b. slightly acidic
   c. slightly basic
   d. very basic

3. A solution having a pH of 9.4 is ________________ than one having a pH of 8.2.

   a. less acidic
   b. more acidic

MOLECULAR MODELS: GETTING THE ANGLE ON CARBON

23

PURPOSE

To study the bonding of carbon compounds.

PERFORMANCE OBJECTIVES

Following this investigation, you should be familiar with representative organic compounds, their bonding, geometric structures, and some of their properties.

BACKGROUND EXPLANATION

In 1857 Kekule, a German chemist, proposed that carbon has four bonds. He also used ball and stick models to represent his theories. By 1870, formulas of carbon compounds were written with four chemical bonds to other atoms. Methane gas, with the formula CH_4, can be drawn as shown here:

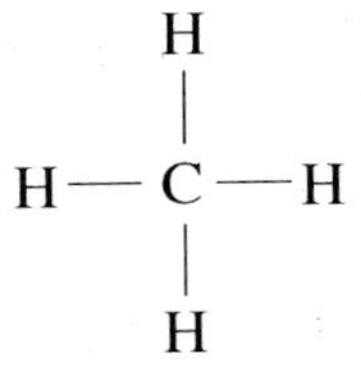

$$
\begin{array}{c}
\text{H} \\
| \\
\text{H} - \text{C} - \text{H} \\
| \\
\text{H}
\end{array}
$$

Substitution of other atoms or groups for hydrogen gives different compounds:

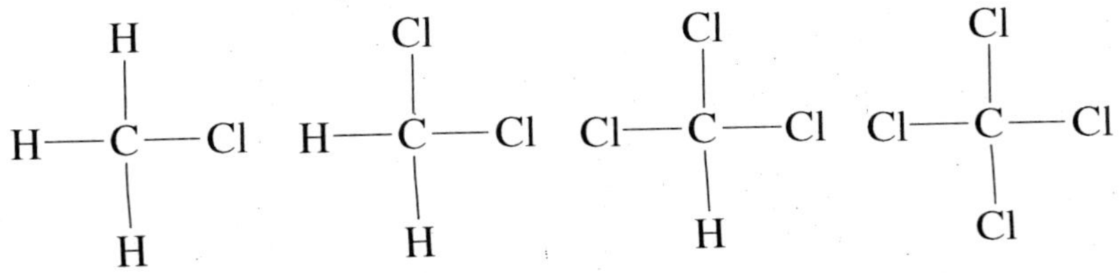

$$
\begin{array}{cccc}
\text{H} & \text{Cl} & \text{Cl} & \text{Cl} \\
| & | & | & | \\
\text{H}-\text{C}-\text{Cl} & \text{H}-\text{C}-\text{Cl} & \text{Cl}-\text{C}-\text{Cl} & \text{Cl}-\text{C}-\text{Cl} \\
| & | & | & | \\
\text{H} & \text{H} & \text{H} & \text{Cl}
\end{array}
$$

Chloromethane   Dichloromethane   Trichloromethane   Carbon
(methyl chloride)   (methylene chloride)   (chloroform)   tetrachloride

It might appear that methylene chloride should have two possible structures:

$$\begin{array}{ccccc}
 & \text{Cl} & & & \text{H} \\
 & | & & & | \\
\text{H}-&\text{C}&-\text{Cl} & \text{and} & \text{Cl}-\text{C}-\text{Cl} \\
 & | & & & | \\
 & \text{H} & & & \text{H}
\end{array}$$

This would be the case if the atoms of the molecule were all in the same plane, but chemical evidence, such as the existence of only one substance with the formula CH_2Cl_2, requires that the four bonds be directed toward the corners of a tetrahedron with the carbon atom at the center.

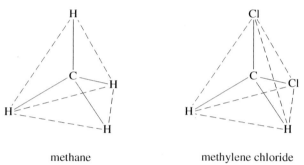

methane   methylene chloride

With the carbon at the center of the tetrahedron, all angles from the center are 109.5° and the methylene chloride has only the one structure shown.

The significance of these chemical bonds between atoms represented by lines became evident soon after the discovery of the electron. A classic paper presented in 1916 by G. N. Lewis proposed that the single bonds shown in structural formulas involves a pair of electrons, and that an atom tends to hold eight outer electrons. The proposal by Lewis, now called the octet rule, can be applied to many atoms.

For some molecules, it is possible to rearrange the atoms to form different structures while satisfying the octet rule. Two different molecules, for example, have the formula C_2H_6O. They are

$$\begin{array}{cccccc}
\text{H} & \text{H} & & & \text{H} & \text{H} \\
| & | & & & | & | \\
\text{H}-\text{C}-&\text{C}-\ddot{\text{O}}-\text{H} & \text{and} & \text{H}-\text{C}-\ddot{\text{O}}-\text{C}-\text{H} \\
| & | & & & | & | \\
\text{H} & \text{H} & & & \text{H} & \text{H}
\end{array}$$

ethyl alcohol   dimethyl ether

These two molecules—called **isomers**—have different chemical and physical properties. Isomerism is very common in organic molecules.

The molecular structure of benzene, C_6H_6, posed a problem to investigators who found that all six hydrogen atoms are equivalent. Kekule proposed that the carbon atoms were in a ring with alternate single and double bonds. The structure could be drawn with double bonds in two alternate positions:

162

These structures are called resonance structures; neither structure is quite accurate. The bond distances in benzene are actually equivalent and are of an intermediate length. If the octet rule is followed, no drawing accurately depicts the bonding. Chemists often use the following structure to represent benzene, which does not behave like typical substances having double bonds.

PROCEDURE

Make ball and stick models for each of the organic compounds. Draw structural formulas and indicate polarity.

Many molecular model kits use the following scheme:

Color	Number of Holes	Atoms represented
Yellow	1	H
Green	1	Cl
Orange	1	Br
Purple	1	I
Red	2	O or S
Black	4	C
Blue	5	N or P

Short sticks represent single bonds to hydrogen. Long sticks are used for all other single bonds. A pair of springs bent to connect two atoms represents a double bond. A triple bond is represented by three springs joining two atoms.

Experiment 23: Organic Models

Compound	Structural Formula	Is the structure polar or nonpolar?
Sample: H_2O water	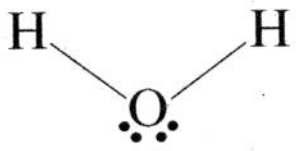 H H $\searrow$ $\swarrow$ $\cdot\ddot{O}\cdot$	polar

1. CH_4
   methane

2. CH_3CH_3
   ethane

3. $CH_3CH_2CH_3$
   propane

4. $CH_3CH_2CH_2CH_3$
   butane

$$\begin{array}{c} CH_3 \\ | \end{array}$$

5. $CH_3—CH—CH_3$
   isobutane (2-methylpropane)

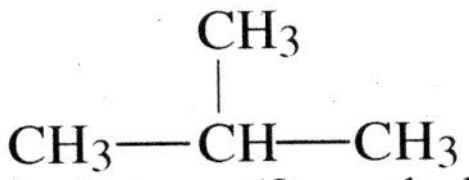

$$\begin{array}{ccc} CH_3 & & CH_3 \\ | & & | \end{array}$$

6. $CH_3—C—CH_2—CH—CH_3$
   $$\begin{array}{c} | \\ CH_3 \end{array}$$

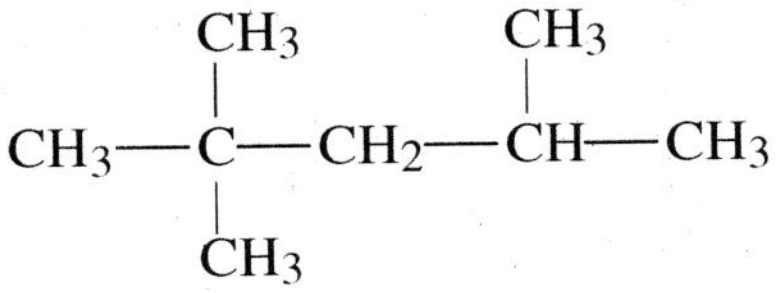

   2,2,4-trimethylpentane (gasoline)

7. CH_3OH
   methyl alcohol
   (methanol)

8. CH_3CH_2OH
   ethyl alcohol
   (ethanol)

9. $CH_3CH_2CH_2OH$
   propyl alcohol
   (1- propanol)

$$\begin{array}{c} CH_3 \\ | \end{array}$$

10. $CH_3—CH—OH$
    isopropyl alcohol
    (2-propanol)

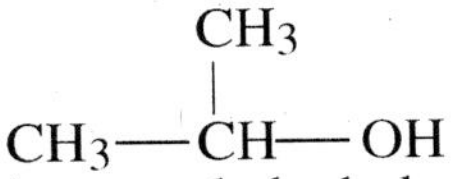

11. The general formula for an alcohol is ROH when R represents any alkyl (carbon) group. How many isomers of propyl alcohol are there? _____

165

12. $CH_3CH_2CH_2CH_2Br$
    butyl bromide
    (1-bromobutane)

13. 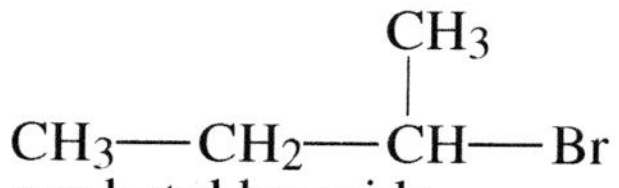
$$CH_3-CH_2-\overset{\overset{\displaystyle CH_3}{|}}{CH}-Br$$
    sec-butyl bromide

14. 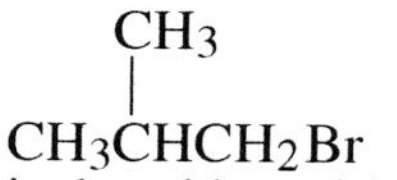
$$CH_3\overset{\overset{\displaystyle CH_3}{|}}{CH}CH_2Br$$
    isobutyl bromide

15. 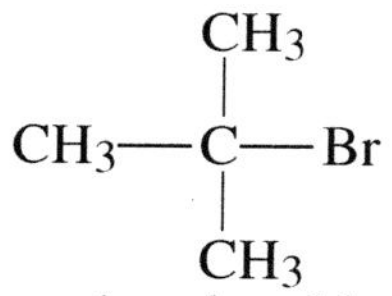
$$CH_3-\overset{\overset{\displaystyle CH_3}{|}}{\underset{\underset{\displaystyle CH_3}{|}}{C}}-Br$$
    tertiary-butyl bromide

16. How many kinds of butyl groups are there?_____

17. C_6H_6
    benzene

    (Use wood sticks for single bonds and two springs for each double bond. Molecules containing the benzene ring are classified as *aromatic*.)

18. CH_3OCH_3
    dimethyl ether

19. What molecule made previously has the same atoms arranged differently, and therefore is an *isomer* of dimethyl ether? _____

20. $CH_3OCH_2CH_3$
    ethyl methyl ether
    (The general formula for an ether is ROR.)

21. $CH_3CH_2OCH_2CH_3$
    diethyl ether
    (Once used for anesthesia, and called ether.)

22. CH_3COOH
    acetic acid
$$CH_3-C\overset{\displaystyle O}{\underset{\displaystyle OH}{\big\backslash}}$$

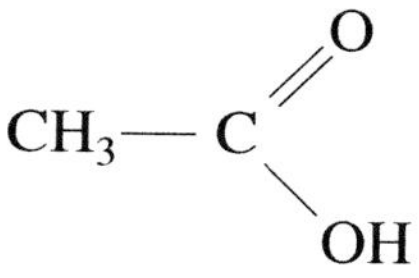

    (The —COOH, called a carboxyl group, is present in all organic acids. Vinegar contains acetic acid.)

23. $CH_3CH_2CH_2COOH$
    butyric acid
    (Butyric acid is present in rancid butter and blue cheese.)

24. $H-\overset{\overset{\displaystyle O}{\|}}{C}-H$
    formaldehyde
    All aldehydes have the general formula $R-\overset{\overset{\displaystyle O}{\|}}{C}-H$.

25. $CH_3-\overset{\overset{\displaystyle O}{\|}}{C}-CH_3$
    propanone
    (acetone)

26. $CH_3-\overset{\overset{\displaystyle O}{\|}}{C}-CH_2CH_3$
    butanone
    (ethyl methyl ketone)
    (All ketones have the general formula $R-\overset{\overset{\displaystyle O}{\|}}{C}-R$.)

27. $CH_3CH_2CH_2\overset{\overset{\displaystyle O}{\|}}{C}-OCH_2CH_3$
    ethyl butanoate, pineapple flavor
    (ethyl butyrate)
    Esters are often used in flavoring.

28. CH_2CH_2
    ethene
    (ethylene)

29. How many bonds must there be between carbon atoms in ethylene?_____

30. $[-CH_2\ CH_2-]_n$

    one unit of polyethylene.
    Join four or more polyethylene units to show a segment of polyethylene.

31.   Cl   H
        \\   /
          C = C
        /   \\
      H   Cl

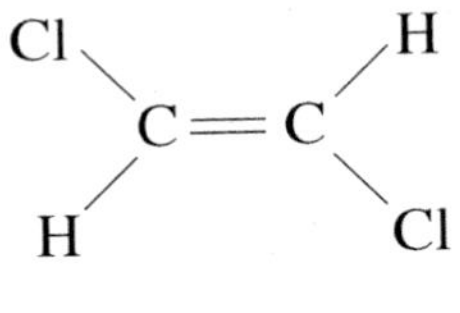

    trans-1,2-dichloroethene

32.   Cl   Cl
        \\   /
          C = C
        /   \\
      H   H

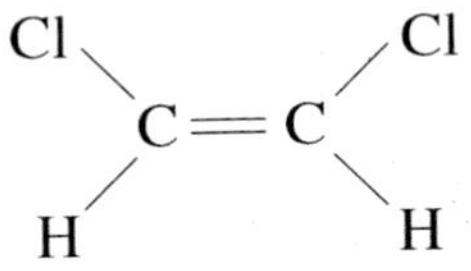

    cis-1,2-dichloroethene

167

33. C_2H_2
   ethyne (acetylene)

34. How many bonds must there be between carbon atoms in acetylene? _____

PRE-LAB or POST-LAB QUIZ ______________________ ______________________
Name Instructor/Section

Experiment 23: PRE-LAB or POST-LAB QUIZ

Multiple choice questions. Circle your answers.

1. In carbon-containing compounds having only single bonds, there are always

   a. 2 bonds in a linear arrangement.
   b. 2 bonds in an angular or bent arrangement.
   c. 3 bonds that form a trigonal pyramidal arrangement.
   d. 4 bonds in a tetrahedral arrangement.

2. Molecules like CH_3CH_2OH and $CH_3—O—CH_3$ that have the same chemical formula, C_2H_6O in this case, are called

   a. isomers.
   b. isotopes.
   c. allotropes.
   d. resonance structures.

3. To represent a double bond with ball-and-stick models, use

   a. 2 short sticks to connect 2 "atoms."
   b. 2 long sticks to connect 2 "atoms."
   c. 2 springs bent to connect 2 "atoms."
   d. 2 springs (not bent) to connect 2 "atoms."

4. The two structures shown here

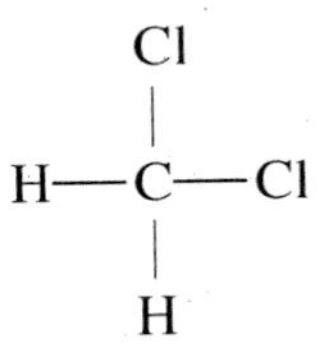

and

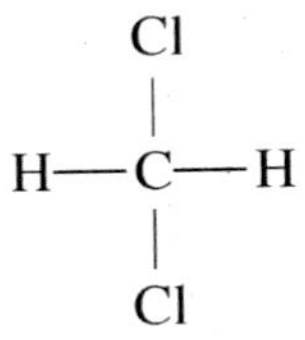

   a. represent the SAME substance.
   b. represent DIFFERENT substances.
   c. do not provide enough information to indicate whether the substances represented are the same or different.

ASPIRIN AND OIL OF WINTERGREEN

PURPOSE

To prepare samples of two esters, aspirin and methyl salicylate.

PERFORMANCE OBJECTIVE

Following this investigation, you should be familiar with several organic chemicals of practical importance.

BACKGROUND INFORMATION

Both aspirin and oil of wintergreen can be prepared from salicylic acid. Salicylic acid derives its name from the Latin class name for the family of willow trees, *salicaceae* (genus *salix*). A crude form of salicylic acid called salicin, extracted from the bark of willow trees, was once used to relieve pain. It can also be extracted from the meadowsweet plant (Latin, *spirea*). Its use in treatment of malaria symptoms was reported in a paper by Edward Stone, a Christian clergyman, to the Royal Society of London in 1763. In search for a substitute for quinine in short supply in the 1800's Charles Gerhardt, a German scientist, discovered acetyl salicylic acid (aspirin) in 1853. The first practical synthesis of aspirin was done by Felix Hoffman of the Bayer firm in Germany in 1893. The Bayer firm named it aspirin. Aspirin is an analgesic (a pain reliever), an antipyretic (a fever reducer) and an anti-inflammatory drug.

Aspirin, acetyl salicylic acid, is an ester of salicylic acid and acetic acid which can be made by refluxing (continuous distillation and condensation) for several hours. It can be made much faster by using acetic anhydride which is more reactive. Either concentrated sulfuric acid or 85% phosphoric acid may be used as an effective catalyst for the reaction.

Oil of wintergreen (methyl salicylate) was first isolated from an extract of the wintergreen plant (Latin, *gaultheria*) in 1843. The wintergreen plant keeps its leaves through the winter; its name is taken from this characteristic. The oil extracted from the plant has a very pleasant odor and when sweetened has a very pleasant taste as you probably already know. Like salicylic acid and acetyl salicylic acid, methyl salicylate is a pain reliever and fever reducer. It is sometimes used in pain relieving lotions and used to relieve toothache. It may be synthesized from salicylic acid and methanol using sulfuric acid as a catalyst:

salicylic acid	methanol	methyl salicylate (oil of wintergreen)	water

Phenol which has the structure and other compounds containing this "phenolic" structure show a purple color when tested with $FeCl_3$. Salicylic acid contains the phenolic group but pure acetyl salicylic acid (aspirin) does not. $FeCl_3$ can be used to test the purity of aspirin.

PROCEDURE: Wear Eye Protection

Part A. Aspirin.

Weigh a suitable container, such as a weighboat or small beaker to the nearest 0.01 g (or better) on a balance. Record this mass. Add approximately 2 g of salicylic acid (powder) and weigh again to the nearest 0.01 g (or better) and record. Put about 200 mL of tap water in a 400 mL beaker and heat to boiling. Pour the salicylic acid into a 125 mL Erlenmeyer flask and add slowly, dropwise, 5 mL of acetic anhydride measured in a 10 mL graduate. Then add 5 drops of concentrated (18 M) sulfuric acid, H_2SO_4. Use caution with concentrated H_2SO_4. Protect your eyes. Wash off any that contacts skin. Heat this flask in boiling water in the 400 mL beaker, inside the hood, for 10 minutes. Add a few drops of distilled (deionized) water to decompose excess acetic anhydride. Then, remove the flask and cool to about room temperature (below 30° C) and add 40 mL of distilled (deionized) water from a graduated cylinder.

Cool the flask in an ice bath to increase crystallization. Chill a small beaker containing about 10 mL of distilled H_2O for later use. If crystals do not form, scratch the inside of the flask with a stirring rod. Place a filter paper in a Buchner funnel on a filter flask attached to an aspirator. Wet the filter paper with distilled water and turn on the suction. Swirl the flask and pour the contents carefully onto the filter paper. Adjust the water flow through the aspirator for maximum vacuum. Wash crystals with about 10 mL of chilled distilled H_2O. Leave water running until crystals are sparkling and dry if you have time. Remove the filter paper with aspirin and dry in open air. Weigh the aspirin in a small beaker to the nearest 0.01 g. Record under DATA. Using the data, calculate the actual yield of aspirin in grams, and the percentage yield of aspirin. Record under RESULTS.

To test the purity of your aspirin add a drop or two of 1% ferric chloride, $FeCl_3$, solution to a few crystals of the aspirin. A yellow color (color of $FeCl_3$) indicates aspirin is pure. A light purple color indicates it is slightly impure. A dark purple indicates more impurity. Record color under RESULTS. If you have or can obtain some commercial aspirin tablets, you may wish to test them. The results may surprise you.

$$\% \text{ yield} = \frac{\text{Mass of aspirin produced}}{\text{Theoretical yield in grams}} \times 100\%$$

Part B. Oil of Wintergreen.

Weigh approximately 1 g of salicylic acid in a suitable container (weighboat for example). Record this mass, but it is not used here for any calculations. Transfer this salicylic acid to a 50 mL Erlenmeyer flask. Add 10 mL of methanol, CH_3OH, and 3 drops of concentrated (18 M) sulfuric acid, H_2SO_4. Use caution with concentrated H_2SO_4, Protect your eyes. Wash off any acid that contacts your skin. Place the flask in a water bath made by using a 400 mL beaker containing about 200 mL of tap water and heat (at boiling) for 15 minutes. Remove the flask using flask tongs and cautiously smell the contents of the flask. Record the odor (or fragrance) under RESULTS. Add a drop of 1% ferric chloride, $FeCl_3$ to the flask to test for any phenolic group. Note any color change. Record these RESULTS.

Name   **Instructor/Section**   **Lab Partner**   **Date**

Experiment 23: Aspirin and Oil of Wintergreen

A. ASPIRIN PREPARATION DATA

1. Mass of container (or weighboat) empty   _____ g

2. Mass of container plus salicylic acid   _____ g

3. Mass of salicylic acid (by subtraction)   _____ g

4. Mass of beaker for aspirin   _____ g

5. Mass of beaker plus aspirin   _____ g

6. Mass of dry aspirin (actual yield)   _____ g

RESULTS

7. Molar mass of salicylic acid, $C_7H_6O_3$ _____ g/mole

8. Molar mass of acetyl salicylic acid, $C_9H_8O_4$ _____ g/mole

9. Theoretical yield of aspirin based on mass
    of salicylic acid used   _____ g

10. .Percentage yield   _____ %

11. Color after $FeCl_3$ test   _____

B. OIL OF WINTERGREEN PREPARATION DATA

12. Mass of container, empty   _____ g

13. Mass of container plus salicylic acid   _____ g

14. Mass of salicylic acid used   _____ g

RESULTS

15. Odor in Erlenmeyer flask   _____

16. Color after $FeCl_3$ test   _____

QUESTIONS

1. Should pure aspirin show a purple color when tested with $FeCl_3$? _____

2. Should pure oil of wintergreen show a purple color with $FeCl_3$? _____

3. The percent by mass may be *over 100%!* Why can this happen?_____

175

Experiment 24: PRE-LAB or POST-LAB QUIZ

Multiple choice questions. Circle your answers.

1. The chemical name for aspirin is _____ .

2. The chemical name for oil of wintergreen is _____ .

3. Both aspirin and oil of wintergreen are classified as

   a. aldehydes.
   b. ketones.
   c. organic acids.
   d. phenols.
   e. esters.

4. One test for the purity of the aspirin is to use a few drops of iron(III) chloride (ferric chloride). If the sample tested turns dark purple when iron(III) chloride is applied, this indicates that the sample

   a. is of very high purity.
   b. is of moderately high purity.
   c. has only a slight impurity.
   d. is not pure.